SMART MATERIALS *and* STRUCTURES TECHNOLOGIES

THE IMPENDING REVOLUTION

An intelligence report on electro-rheological fluids, memory metals, piezo-electric actuators and sensors, embedded fiber optics sensing systems and smart ultra-advanced composites

This intelligence report was prepared by HY-TECH INTERNATIONAL, Okemos, Michigan

The principal authors are

DR. BRIAN S. THOMPSON
DR. MUKESH V. GANDHI

Smart Materials and Structures Technologies
a **TECHNOMIC**® publication

Published in the Western Hemisphere by
Technomic Publishing Company, Inc.
851 New Holland Avenue
Box 3535
Lancaster, Pennsylvania 17604 U.S.A.

Distributed in the Rest of the World by
Technomic Publishing AG

Printed in the United States of America
10 9 8 7 6 5 4 3 2

Main entry under title:
Smart Materials and Structures Technologies: The Impending Revolution

A Technomic Publishing Company book
Bibliography: p.

Library of Congress Card No. 89-51657
ISBN No. 87762-720-7

TABLE OF CONTENTS

INTRODUCTION

Materials technology has had such a profound impact on the evolutionary status of Homo sapiens that historians have characterized several distinct time-periods in the evolution of human civilization by terms such as the Stone Age, the Bronze Age and the Iron Age. The classical quest for superior products in the commercial, industrial and military segments of the economy, which were responsible for these eras, is very much in evidence today as the time-scale approaches the year 2000. Humankind's insatiable demand for materials with superior performance characteristics has already precipitated the age of synthetic composite materials. Whereas the first record of this embryonic technology dates back to 1446 B.C., the field has lain dormant until the burgeoning activities in the latter half of the 20th century. This document reports on one of these activities which is focused on the development of innovative classes of ultra-advanced smart materials and structures technologies.

Whereas the materials technologies inspired by mankind's quest to conquer the last frontier of space have had a profound effect on the technology base, the dawn of the 21st century will witness the revolution in smart materials and structures technologies which is an outgrowth of the technologies spawned by the aerospace industries. This revolution will provide the nervous system, the brains, and the muscles for the existing advanced materials and structures which are a mere skeleton at this time compared with the anatomy perceived in the not-too-distant future. This quantum jump in materials technology will revolutionize the future in ways far more dramatic than the way the chip has catalyzed the evolution of the electronics industry.

This intelligent report has been compiled by the professional staff of Hy-Tech International, which includes research scientists, business management consultants and economists. Since the professional staff at Hy-Tech International serves as consultants to various government agencies, Fortune 500 companies and major high-technology-oriented companies worldwide, this report provides only a qualitative technological profile of the activities of these companies. The professional staff at Hy-Tech International are at your disposal to provide technical and management services which can catalyse the evolution of a competitive edge for your company in this rapidly evolving field.

Technomic Publishing Company is a leading producer of technical and marketing reports, books, and journals that deal with key technologies of today and tomorrow. The employees and associates of Technomic Publishing Company have many years of experience in these technological markets and the resultant contacts.

The present venture in Smart Materials with Hy-Tech International is another step in the production of valuable information in a new and emerging field of technology. Technomic Publishing Company believes that the conclusions given in this report constitute an accurate picture of the status of Smart Materials today.

1 EXECUTIVE SUMMARY

The insatiable demand in the international marketplace for high performance structural and mechanical systems for the aerospace, defense, and advanced manufacturing industries has triggered the evolution of advanced composite materials. With traditional advanced composite materials the optionization strategies result in an optimal design that is passive in nature and cannot respond to unstructured environments. An optimally designed product in a traditional advanced composite material is passive in the sense that it cannot actually respond to changes in various conditions of its operation. The elasto-dynamic response of such a product is suboptimal for all of the service conditions except those for which it is optimally designed. Since these elasto-dynamic phenomena manifest themselves in practically all applications there is a significant need for the evolution of a new class of advanced composite materials whose elasto-dynamic response can be optimally tailored in real time in order to significantly enhance the performance of structural and mechanical systems under a diverse range of operating conditions.

Smart materials and structures are those that exploit the muscular characteristics offered by electro-rheological fluids, piezo-electric materials, and shape memory metals, and also the sensing characteristics offered by fiber optic systems, for example. These smart materials and structures have the unique capabilities to adapt to changes and external stimuli capitalizing on the intrinsic intelligence embedded in them. Electro-rheological fluids are suspensions of fine particles typically polymeric or metallic in non-conducting liquids such as silicone oils and heptane. These suspensions can be changed reversibly from a liquid-like state to a solid-like state from the application of an electrical field.

Shape memory alloys are typically copper alloy systems and nickel titanium alloys that change their shape and properties reversibly when heated. Subsequent cooling of the metal permits them to recover their original shape and properties; therefore, these materials are also called shape memory alloys.

Piezo-electric materials are materials that generate a charge in response to a mechanical deformation or alternately they provide mechanical strain when an electric field is applied across them. This capability permits piezo-electric materials to be employed either as activators or sensors in the development of smart structures.

Fiber-optics is a rather immature sensing and data transmission technology in which light is transmitted through fibers. The technology offers many advantages over conventional sensor technologies and it may be employed to measure a broad range of phenomena.

Smart materials do have major potential advantages; however, they also have technical deficiencies. The technical deficiencies of electro-rheological fluids are related to the following issues:

1. Stability
2. Thermal effects
3. Electrode separation
4. Certification
5. Interaction
6. Continuum applications
7. Control issues

Piezo-electric material deficiencies are related to the following issues:

1. Strength reduction
2. Thermal effects
3. Certification
4. Interaction
5. Continuum applications
6. Control issues

The deficiencies of shape memory alloys are related to the following issues:

1. Slow response times
2. Hysteresis
3. Energy
4. Temperature constraints
5. Stress concentration
6. Certification
7. Interaction
8. Continuing applications
9. Control issues

Fiber-optic sensors have the following technical deficiencies:

1. Integration issues
2. Sensitivity

Smart materials and structures technologies that are presented in this report will have a tremendous impact on reshaping the technological and economic base of the international business environment during the next two decades. It is expected that the impending revolution of smart technologies will impact every segment of the world marketplace. It is expected that if the smart technologies field matures, individual technologies will be integrated to yield hybrid smart technologies that will permit the stringent performance specifications of various smart parts and subsystems to be adequately satisfied. Therefore, the complete impact of smart materials is impossible to forecast or predict at this point in time. However, examples of the marketplace and the future market projections can be given.

The industries that are most likely to utilize smart technologies are:

- automotive and transportation
- aerospace
- defense
- biomedical devices
- advanced manufacturing
- robotics and industrial machinery
- various consumer and sporting good products
- high precision instruments
- civil engineering projects—highways, buildings, and bridges

It is estimated that the total market of smart materials and structures will be at least 65,000 million dollars (U.S.) by the year 2010.

The race in the smart materials arena has already started in the aerospace sector of the economy, discrete areas of the automotive industries, and some smoke-stack industries. Furthermore, it is believed that the first applications of these smart materials technologies will be in the aerospace industries with viable prototypes being developed by 1992. These research and development activities will initially be fueled by IRAD monies to the principal U.S. defense contractors who stand to reap a bountiful harvest in both profits and market share through the successful development of these embryonic technologies. Specific programs that would benefit from quantum leaps in smart materials technologies include the Strategic Defense Initiative, the F-15, the National Aerospace Plane, the NASA Space Station, the B2 bomber, and numerous weapon systems incorporating stealth technology.

The aerospace industries and the defense industries will be the first to significantly fully capitalize on these smart technologies. It is expected that the ultimate applications of these smart technologies in the industries mentioned will probably not be realized until after the year 2000.

There are commercial activities around the world with at least 20 of the leading industrial companies doing work on smart technologies and structures. These companies are at varying stages of development, but it is expected that most of them will be continuing their efforts in these areas. Companies in the United States who are very active in these areas include:

- American Cyanamid
- Boeing
- Chrysler
- DuPont
- FMC
- Ford
- General Motors
- Hercules
- Lockheed
- Lord
- McDonnel Douglas
- Raytheon
- Rockwell
- Sikorsky
- United Technologies

In Japan the following companies are also very active in this area:

- Toyota
- NEC
- Toshiba
- Mitsubishi Heavy Metal Industries
- Mitsubishi Electric
- Kawasaki Heavy Industries

2

CONCLUSIONS AND RECOMMENDATIONS

The impending revolution in smart materials and structures technologies must be carefully evaluated at the highest level in every major company. Optimal positioning in the marketplace could mean billions in profits and big jumps in market share. A passive strategy will result only in the loss of existing markets and ultimate failure since most of the major players on the international scene are already scrambling to establish a distinct edge in this area.

(a) Every major high-technology-oriented company *must* carefully evaluate the creation of new markets and also the impact of the impending revolution in smart materials and structures technologies on its product lines. Most of the major international players are already in the game, and no major company should position itself to be left standing at the starting line.

(b) A task force must be established at the very highest levels in the organization to identify the relevant groups and technical programs which could contribute most to the company's endeavors in the field of smart materials and structures. The task force must be of an interdisciplinary nature and consist of diverse technical people, human resource specialists, management consultants, marketing strategists and economists. It is strongly recommended that outside consultants should be exploited at this task-force level to either fill the voids in the profile and/or integrate the discrete pockets of expertise which constitute the group. A comprehensive listing of potential consultants in the area is presented in Chapter 9.

(c) Human resources will be the single most important impediment to the establishment of a viable group in the area of smart materials and structures for any company. Due to the embryonic nature of these technological disciplines, no major university has focused graduate-level programs in this area. This lack of skilled technical and management personnel in the area of smart materials and structures is further exacerbated due to the interdisciplinary nature of the field. Therefore, consultants in the field should be retained to train and educate existing personnel. A comprehensive listing of potential consultants in the area is presented in Chapter 9.

(d) It is estimated that a minimum investment of 1.0 million U.S. dollars per year is necessary to initiate small-scale laboratory proof-of-concept studies focused on one or two specific technologies in the field of smart materials and structures.

(e) A sustained R & D program for three to five years in the field of smart materials and structures is a necessary prerequisite for launching a viable activity in the smart materials marketplace.

(f) The possibility of spinning off a small business subsidiary focused on smart materials should not be excluded from consideration. Such a subsidiary would offer a forum for focused efforts in the area without distraction from day-to-day preoccupations and pressures. Furthermore, such a company would provide team members with the pride and stimulation of entrepreneurship, in addition to pragmatic advantages such as qualification for state- and federally-funded small business innovation research programs.

3 GENERAL OVERVIEW

Smart materials and structures are a new class of materials and structures that have self-inspection and inherent adaptive capabilities. Typically, the ingredients of a smart structure include:

- a structural material
- a network of sensors
- a network of actuators
- microprocessor-based computational capabilities
- real-time control capabilities

In this novel class of materials and structures, the network of actuators provide the muscles to make things happen, the network of sensors provide the nervous system to monitor and communicate the external stimuli, the structural materials provide the skeleton, and the microprocessor-based computational capabilities provide the brains which ensure the optimal performance of the overall system. Smart materials and structures that exploit the muscular characteristics offered by electro-rheological fluids, piezo-electric materials, and shape-memory metals and also the sensing characteristics offered by fiber optic systems, for example, are schematically presented in Figures 3.1 through 3.4.

Smart materials and structures have the unique capability to adapt to changes in external stimuli by capitalizing on the intrinsic intelligence embedded in them. Since most materials and structures operate under variable service conditions and in unstructured environments, there should be a tremendous demand in the international marketplace for high-performance smart structural and mechanical systems that can respond to external stimuli in an autonomous sense.

For instance, the vibrational characteristics of a helicopter rotor fabricated in an advanced composite material are clearly dependent upon several factors such as rotational speed, aerodynamic loading, payload and ambient hygrothermal conditions. An optimally-tailored rotor designed in a traditional advanced composite material is *passive* in the sense that it cannot actively respond to changes in the environment. It is clearly evident, therefore, that the elastodynamic response of

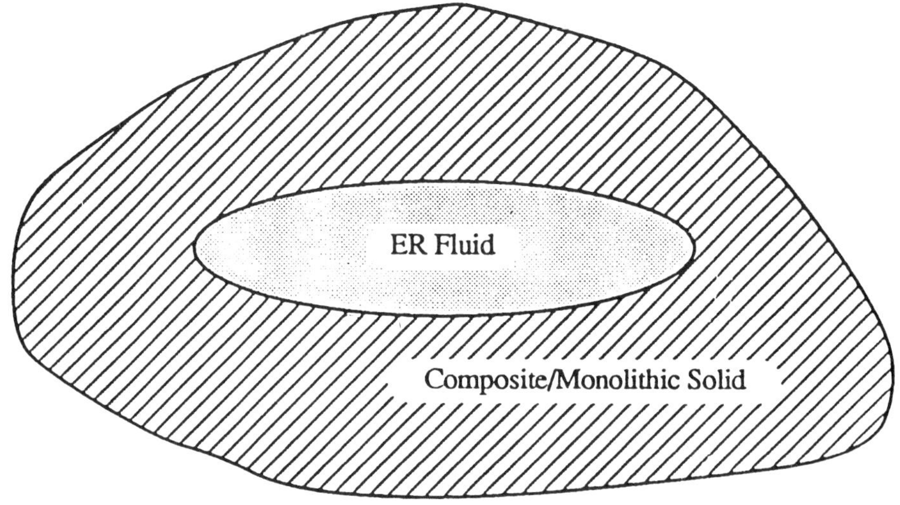

ER Fluids	Structural Materials
Fluid 1	Aluminum Alloys
Fluid 2	Carbon Steels
Fluid 3	Magnesium Alloys
	Graphite/Epoxy Materials
	Glass/Epoxy Materials
	Hybrid Composite Materials

FIGURE 3.1 Electro-rheological (ER)-based Smart Structure.

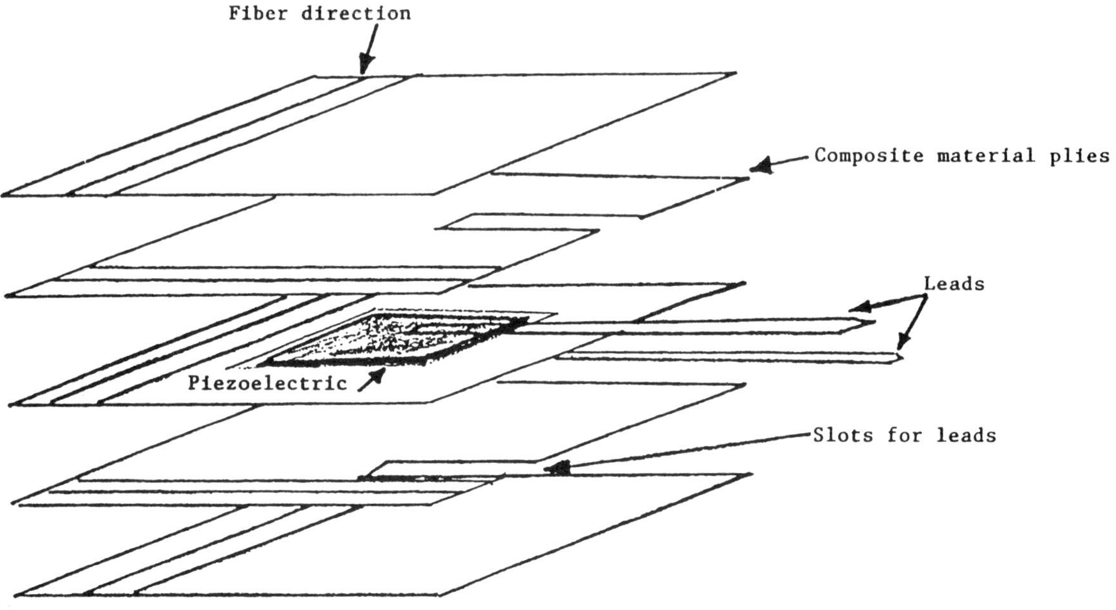

FIGURE 3.2 Piezoelectric-based Smart Structure.

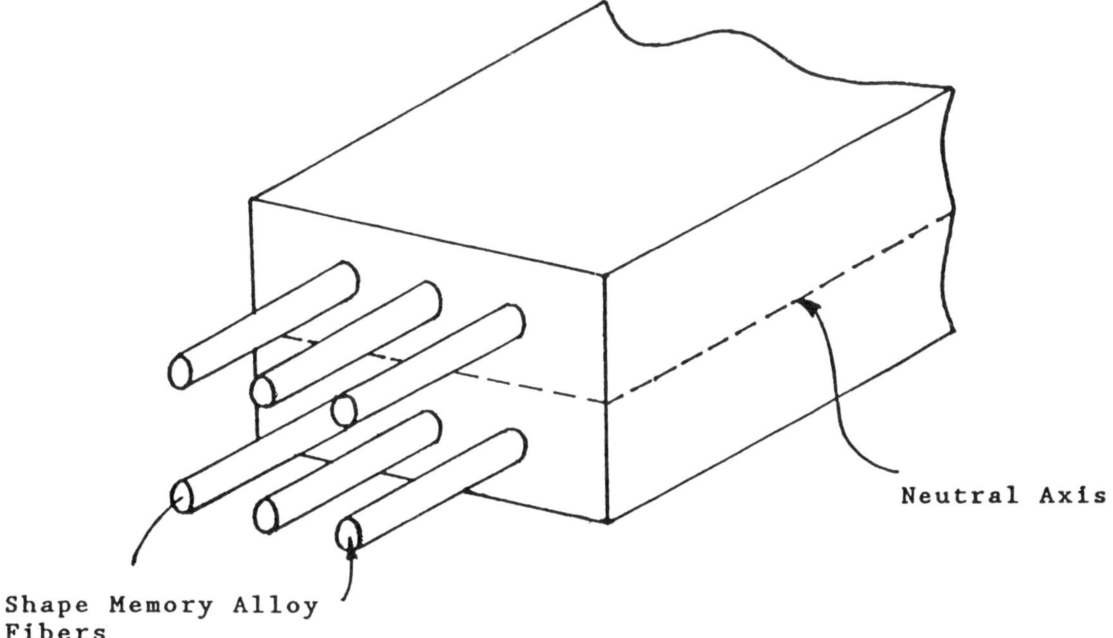

FIGURE 3.3 Shape memory alloy-based Smart Structure.

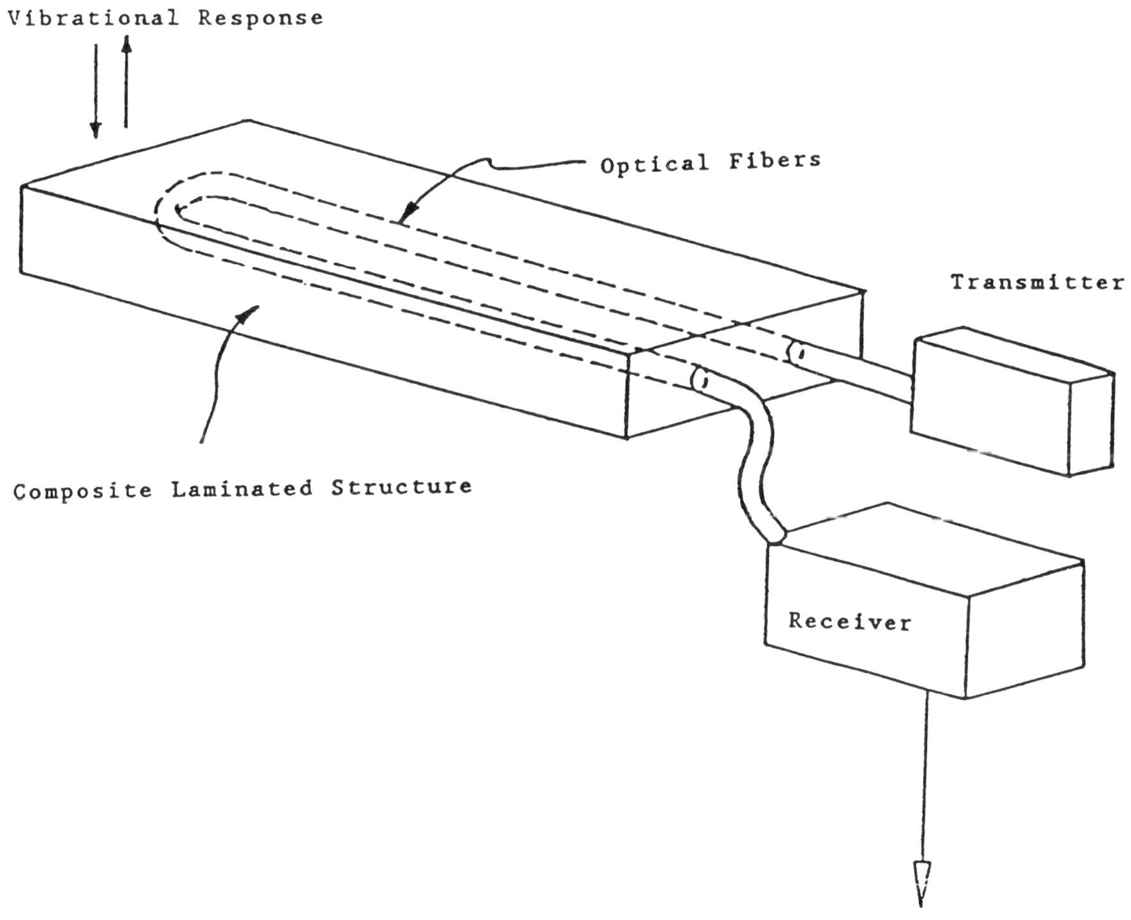

FIGURE 3.4 Fiber optic sensing for Smart Structures.

the rotor is sub-optimal for all service conditions except the one for which the rotor was "optimally designed." In sharp contrast to this undesirable scenario, if the rotors were fabricated in one of the smart ultra-advanced composite materials discussed in this intelligence report, then the performance of the rotor could be dynamically-tuned to ensure optimal performance under all service conditions and in all unstructured environments. Therefore, the technology of smart materials and structures represents a revolutionary change in the technology-base relative to the current generation of advanced composite materials in the marketplace at this time.

A helicopter rotor designed and fabricated in a smart material would have the capability to detect dynamic strains due to changes in payload, wind gusts, temperature and moisture conditions, and battle-field damage, for example. These changes in the vibrational response characteristics of the rotor, changes in the environment, and the qualitative and quantitative measures of damage would be measured by the sensing network and communicated to the appropriate microprocessor, which is the brain of the system. Almost instantaneously, the microprocessor would activate the actuators embedded in the smart structure, which are the muscles of the system, in order to initiate appropriate corrections that would then offset the undesirable effects of the external stimuli on the performance of the rotor system. This activity might involve redistributing loads around highly stressed regions of the rotor structure in order to control the damage sustained by this crucial rotating assembly and ensure the survivability of the helicopter in a hostile battlefield environment, for example.

The most significant impact of these smart materials and structures technologies will be in the following sectors of the economy:

- automotive and transportation industries
- aerospace industry
- defense industry
- biomedical
- advanced manufacturing, robotics and industrial machinery
- consumer products and sporting goods
- high precision instruments, printed circuit boards and electronic packaging
- highways, buildings and bridges

A comprehensive listing of typical applications in these sectors of the economy is presented in Chapter 6.

Ideally, smart materials and structures should demonstrate the following characteristics:

- the ability to respond almost instantaneously to changes in external stimuli
- an inherent ability to interface with modern microprocessors and solid-state electronics
- the inherent ability to integrate modern control systems

These versatile characteristics would provide designers, for the first time, with a unique capability to synthesize ultra-advanced smart composite structures, whose continuum vibrational response can be actively controlled in real time. This class of innovative smart materials derives its intelligence from the merger of sensors built into the finite element control segments of the ultra-advanced composite (or monolithic) material continuum, microprocessors, and dynamically-tunable actuators and/or sensors as shown in Figure 3.5.

The sensors monitor the vibrational behavior of the ultra-advanced, smart, composite structure, and the signals from the sensors are fed to the appropriate microprocessor that evaluates these inputs prior to determining an appropriate control strategy in order to synthesize the desired vibrational response characteristics. This synthesis process is typically accomplished by controlling the mechanical characteristics of desired finite element segments associated with the particular sensors. This change in the characteristics in a typical finite element control segment, in turn, alters the global mass, stiffness, and energy-dissipation characteristics of the ultra-advanced, smart, composite structure in order to achieve the desired vibrational response.

Several sensing technologies may be employed with this class of smart materials and structures. For example, intelligent sensors may be employed by integrating the sensing function and the associated electronic-data-processing function on a

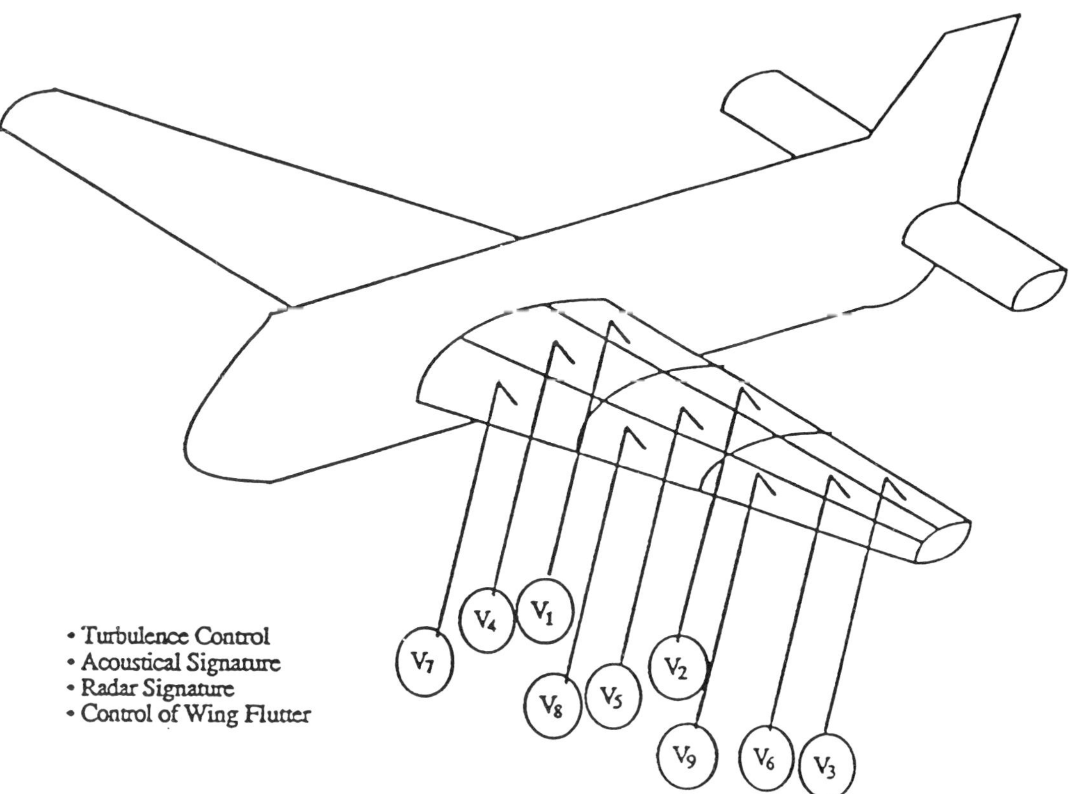

FIGURE 3.5 Finite element control segments of a Smart Aircraft Wing.

single I.C. chip. The current state-of-the-art practice is to embed piezo-electric sensors or fiber-optic sensors and data-links in this class of smart structures. This class of sensors can be manufactured small enough and with high enough strength to avoid degrading the structural integrity of the material.

The most promising dynamically-tunable, actuator technologies are electro-rheological fluids, shape-memory metals and piezo-electric materials. These actuator systems have diverse characteristics, advantages and disadvantages; however, they offer significant capabilities to perform a diverse range of tasks. By judicious selection, the smart materials designer can synthesize numerous classes of hybrid actuation systems from these three principal classes of actuators to satisfy a very broad range of performance specifications for smart systems.

The commercial and economic significance of these smart-materials technologies will be evident once these materials are adequately modeled so that design engineers can capitalize on their unique, adaptive capabilities.

A brief review of technologies that are strategic in the development of smart materials and structures is presented to provide a succinct briefing on each of the principal technologies required for deploying the new generation of smart materials and structures described in this document.

Electro-Rheological Fluids

Electro-rheological fluids are suspensions of fine particles, typically polymeric or metallic, in non-conducting liquids, such as silicone oils or heptane. These suspensions can be changed reversibly from a liquid-like state to a solid-like state upon the application of an electrical field with a typical field strength of 2 kV/mm. The change in phase is accomplished almost instantaneously in approximately 0.001 second to 0.0001 second depending upon the chemical composition of the suspension. Since the current is of the order of 0.001 Amperes to 0.00001 Amperes, the power consumption of this actuator system is low.

Electro-Rheological Fluids: Typical Applications

Electro-rheological (ER) fluids have traditionally been employed in discrete devices associated with hydraulic circuitry. Typical products include electrically-controlled valves for regulating the flow of ER fluids and energy dissipation devices. Automotive products featuring ER fluids include engine mounts and hydraulic clutches.

Electro-Rheological Fluids: Recent Developments

During the past five years there have been two significant developments in the field of ER fluids technology:

(a) The first development is a waterless ER fluid by Filisko and Armstrong at the

University of Michigan and a substantially waterless fluid by Block at Cranfield Institute of Technology in England. These new fluids will operate over broader temperature ranges and at higher temperatures then before, thereby permitting the commercialization of a new generation of ER-based products whose development has been hindered by the previous generation of hydrous fluids.

(b) The second significant development in ER-based systems is the application of a patent by Gandhi and Thompson at Michigan State University for a new class of ultra-advanced composite materials containing voids filled with ER fluids. By controlling the electrical field imposed upon these fluid domains, the mass, stiffness, and the energy dissipation characteristics of the ultra-advanced composite material can also be controlled. These characteristics, in turn, govern the natural frequencies and dissipative behavior of the material. This breakthrough will have a very significant impact on the next generation of smart skins, helicopter rotors and stealth systems, for example.

Shape Memory Alloys

Shape memory alloys (SMA) are typically copper alloy systems and Ni-Ti alloys that change their shape and properties reversibly when heated. Subsequent cooling of the metal permits them to recover their original shape and properties; therefore, these materials are also called shape memory metals. The original shape is typically recovered due to a reversed transformation of the deformed martensitic phase to the higher temperature austenitic phase. The most popular shape memory alloy in use today is Nitinol, which is an alloy of nickel and titanium.

Shape Memory Alloys: Typical Applications

Shape memory alloys are typically sold as actuator systems in the form of wire and thin strip products. The number of applications increased significantly after 1965 when Ni Ti was developed. Pipe couplings, fasteners and clamps of various types have been developed in addition to SMA actuators for robots and mechanisms. Other applications include SMA heat engines, commercial heating units and cooling systems, switches, orthopedics, orthodontics, micropumps for artificial organs and components in medical equipment.

Shape Memory Alloys: Recent Developments

Shape memory alloy applications are burgeoning as the knowledge-base in this field, which is currently dominated by the Japanese, is transferred to other disciplines. For example, wires of SMA have recently been embedded in advanced composite materials in order to develop a new class of smart structures. The application of heat to these embedded SMA actuators permits the structure to respond in a controlled manner to the external stimulus permitting an initially straight beam to be changed to a curved beam.

Piezo-Electric Materials

Piezo-electric materials are materials that generate a charge in response to a mechanical deformation, or alternatively they provide mechanical strain when an electric field is applied across them. This capability permits piezo-electric materials to be employed either as actuators or sensors in the development of smart structures. Piezo-electric sensors and actuators for smart materials applications are fabricated in polymeric materials such as polyvinyldene fluoride (PVDF or PVF_2). Piezo-electric materials have very high mechanical strength and high sensitivity to changes in pressure. Typically, these sensors are used for tactile sensing, temperature sensing and strain sensing, for example.

Piezo-Electric Materials: Typical Applications

Piezo-electric sensors have been employed as tactile sensors to identify objects with a sensitivity sufficient enough to identify the Braille alphabet and various grades of sandpaper, for example. They have also been used as skin-like sensors to mimic the pressure and temperature sensing capabilities of human skin. Furthermore, they have been employed in high-accuracy situations involving the monitoring of mechanical vibrations. In fact, many of the premier accelerometer manufacturers employ piezo-electric crystals in the heart of their products.

Piezo-Electric Materials: Recent Developments

While piezo-electric materials have been extensively employed to exploit the ability of these materials to generate a charge in response to a mechanical excitation, it is only in recent years that engineers have begun to exploit the ability of these materials to develop a mechanical strain in response to the application of an electric field across them.

Recent work has focused on the development of smart structures for aerospace applications featuring numerous piezo-electric actuators and sensors that are employed to control the vibrational behavior of these large flexible structures. Combined theoretical and laboratory studies have been responsible for the development of methodologies that integrate control strategies with the mechanics of the structural behavior.

Fiber Optic Sensing Systems

Fiber optics is a relatively immature sensing and data-transmission technology in which light is transmitted through fibers typically 0.005 inch in diameter. The technology offers many advantages over conventional sensing technologies; it may be employed to measure a broad range of phenomena. The kernel of the technology is to modify and then measure the change in the characteristics of the light beam transmitted through the optical fiber.

Fiber Optic Sensing Systems: Typical Applications

There are over 75 different types of fiber optic sensors for discrete applications involving the measurement of both discrete and field effects. Typical applications range from the measurement of displacements, forces and torques to acoustical and magnetic fields.

Fiber Optic Sensing Systems: Recent Developments

In the context of smart structures, fiber optic sensing systems are now being embedded in advanced composite materials in prototype demonstrations. This research thrust is an ideal application for this sensing technology since the optical fibers are of a similar radial dimension to the structural fibers of graphite or aramid. Moreover, this sensing system permits the manufacture of the part to be monitored, the health of the part to be monitored during service, and the impending failure of the part to be predicted. This application has motivated the optical fiber manufacturers to develop fibers that can withstand the 300°F curing temperatures associated with the manufacture of the smart structural material.

Vibration-Control Strategies

An essential ingredient in the implementation and development of smart materials and structures is the control of the inherent characteristics of the smart structure. Thus, the goal is to actively control the mass, stiffness and energy dissipation characteristics of the structure that in turn govern the vibrational response. Viscoelastic coatings have been traditionally employed as a passive control measure. Other passive control measures include the deployment of advanced composite materials that permit the design engineer to fabricate structural members with high specific stiffnesses, strengths, and energy-dissipation characteristics that are crucial in numerous automotive, aerospace and defense applications.

The classical block diagram for an active strategy is presented in Figure 3.6 where a feedback loop is employed in conjunction with sensors and an adaptive or optimal control algorithm to generate prescribed discrete excitation forces that, when combined with external excitation imposed on the plant, result in the development of acceptable performance characteristics. These technologies are typically employed to control the elastodynamic behavior of robotic devices and large space structures.

This theme of discrete active vibration control has been superseded by the emergence of embryonic technologies dedicated to the distributed-actuator-active control of structures fabricated in monolithic materials, using thermal gradients and piezo-electric effects.

Recently, a new generation of multi-functional, dynamically-tunable, ultra-advanced, intelligent materials has been proposed that capitalize on the

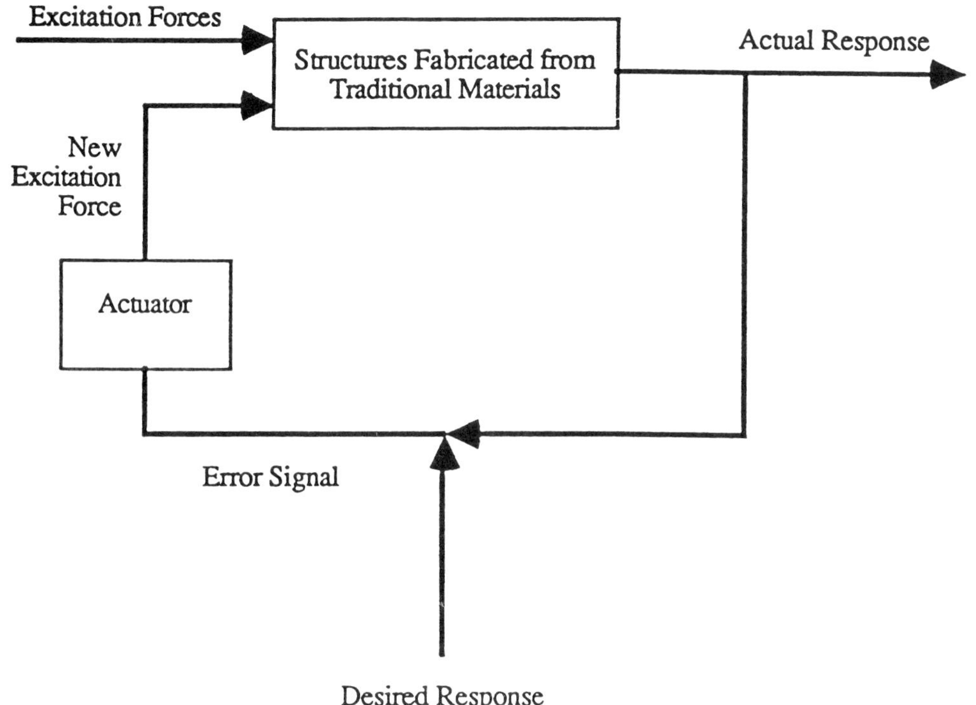

FIGURE 3.6 Traditional control strategies.

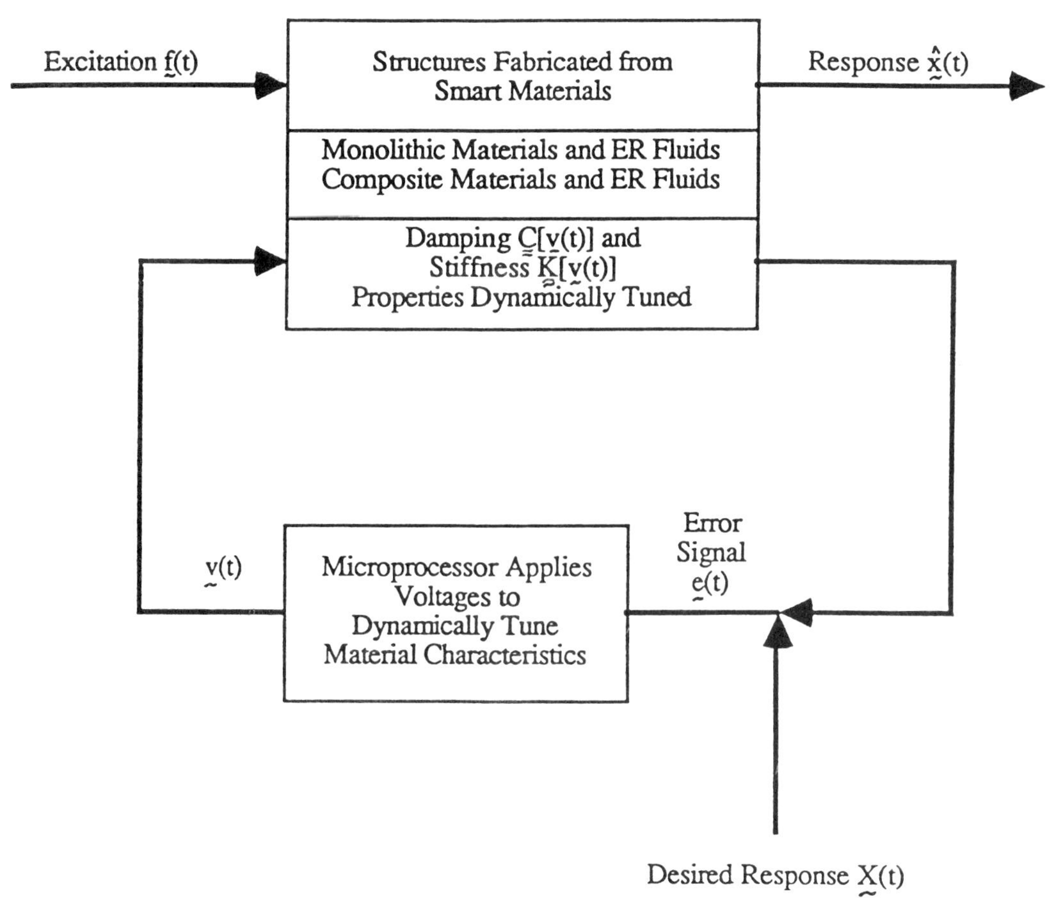

FIGURE 3.7 Control strategies for Smart Materials and Structures.

strengths of active continuum control strategies using distributed actuators and also the freedom offered by advanced composite materials to optimally synthesize the global properties of the materials. This innovative class of potentially revolutionary materials derive their unique qualities from the integration of advanced composite materials technologies, the embedment of smart sensors and fiber-optic data links within the material, and the interface of the structural components of the material with embedded actuators.

A synopsis of the proposed philosophy is presented in Figure 3.7 using a controls format in which a typical smart structure contains actively-controllable electro-rheological fluids that dramatically change their rheological characteristics when subjected to an electrical field. The operating procedure for actively controlling the elastodynamic response involves measuring the actual elastodynamic response of the structure and comparing it with the desired response in order to generate an error signal that is then fed to a microprocessor that ascertains the characteristics of the voltages to be imposed upon the ER fluid domains in that region of the structure in order to develop the desired response characteristics. Thus, in sharp contrast to the conventional scheme presented in Figure 3.6 in which the plant is time-invariant and attention focuses on changing the plant input, the scheme presented in Figure 3.7 focuses upon continuously changing the properties of the plant in real-time. The technological and economic impact of the smart technologies discussed herein is presented in Chapters 6 and 7.

TECHNOLOGICAL STATE-OF-THE-ART

The insatiable demand in the international marketplace for high-performance structural and mechanical systems for the aerospace, defense, and advanced manufacturing industries has triggered the evolution of advanced composite materials. These diverse high-performance applications have mandated that designers tailor the materials and the material microstructural characteristics in order to provide optimal performance of the mechanical and structural systems under various service conditions and unstructured environments. Designers have responded to these challenges by developing optimal design methodologies for a broad class of composite materials. Typically, these methodologies have been employed to optimally select the individual constituents, their micromechanical characteristics, and the spatial distribution of these constituents in order to synthesize a viable structure/subsystem with the desired mass, stiffness, and damping properties.

With traditional advanced composite materials, the optimization strategies result in an optimal design that is passive in nature and cannot respond to unstructured environments. For instance, the vibrational characteristics of a helicopter rotor fabricated in an advanced composite material are clearly dependent upon several factors such as the rotational speed, aerodynamic loading, payload and the ambient hygrothermal environment. An optimally-tailored rotor designed in a traditional advanced composite material is passive in the sense that it cannot actively respond to changes in the rotor speed, aerodynamic loading, etc. It is clearly evident, therefore, that the elastodynamic response of the rotor is suboptimal for all service conditions except the one for which the rotor was optimally designed.

Since elastodynamic phenomena manifest themselves in practically all applications such as submarines, machine tools, aerospace environments, high-speed machinery and robotics, clearly there is a significant need for the evolution of a new class of advanced composite materials whose elastodynamic response can be optimally tailored in real-time in order to significantly enhance the performance of structural and mechanical systems under diverse operating conditions. A methodology for synthesizing this class of smart materials and structures is presented in Figure 4.1.

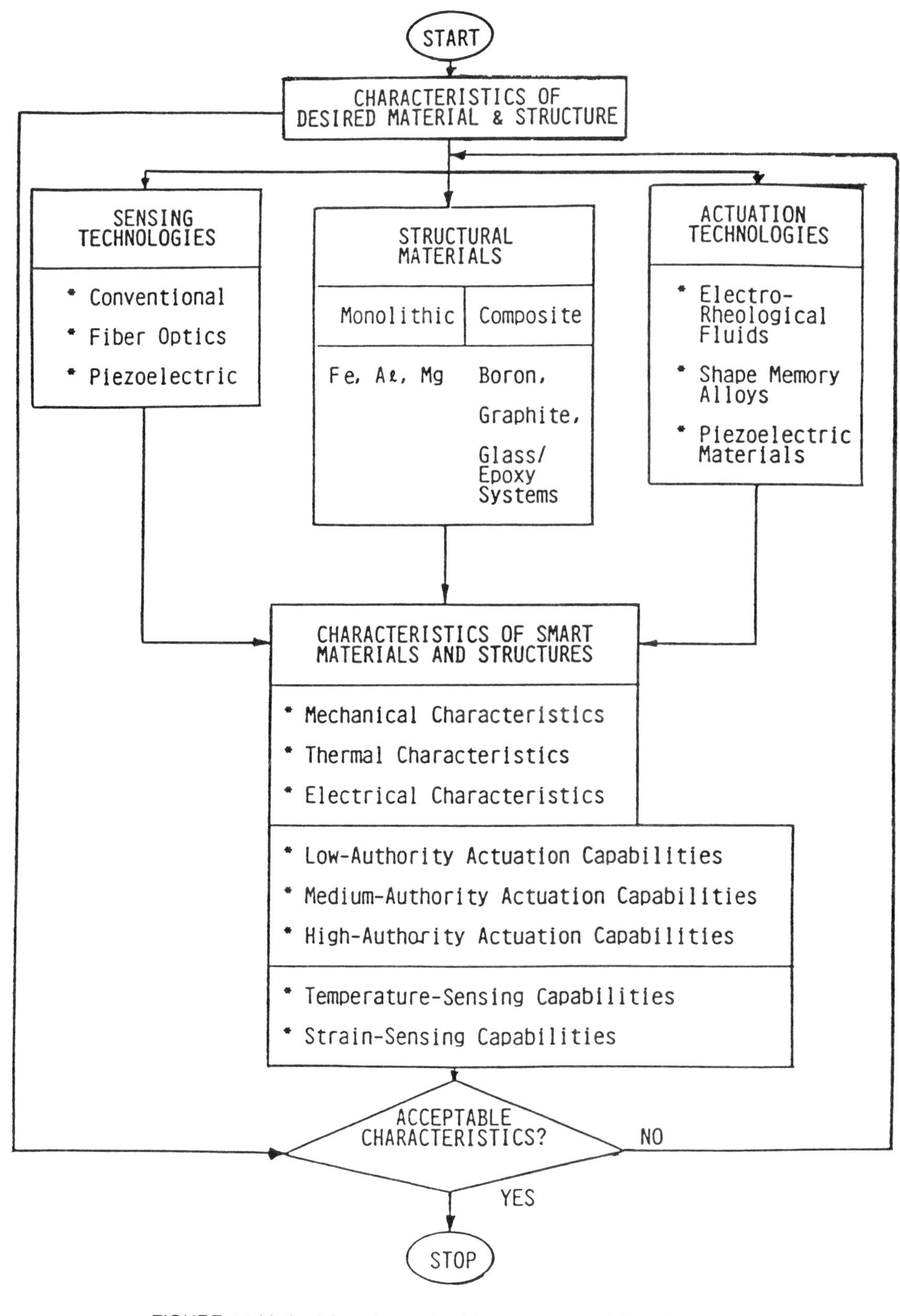

FIGURE 4.1 Methodology for synthesizing smart materials and structures.

4.1 Electro-Rheological Fluids

Background on Electro-Rheological Fluids

Electro-rheological (ER) fluids are typically suspensions of micron-sized hydrophilic particles suspended in suitable hydrophobic carrier liquids, which undergo significant instantaneous reversible changes in material characteristics when subjected to electrostatic or electrodynamic potentials. The most significant change in the material characteristics of an ER fluid is associated with the bulk viscosity of the suspension, which varies dramatically upon applying an electrical field to the fluid. The tailoring of this rheological property by the imposition of a suitable electrical potential can be usefully exploited in vibration-suppression applications.

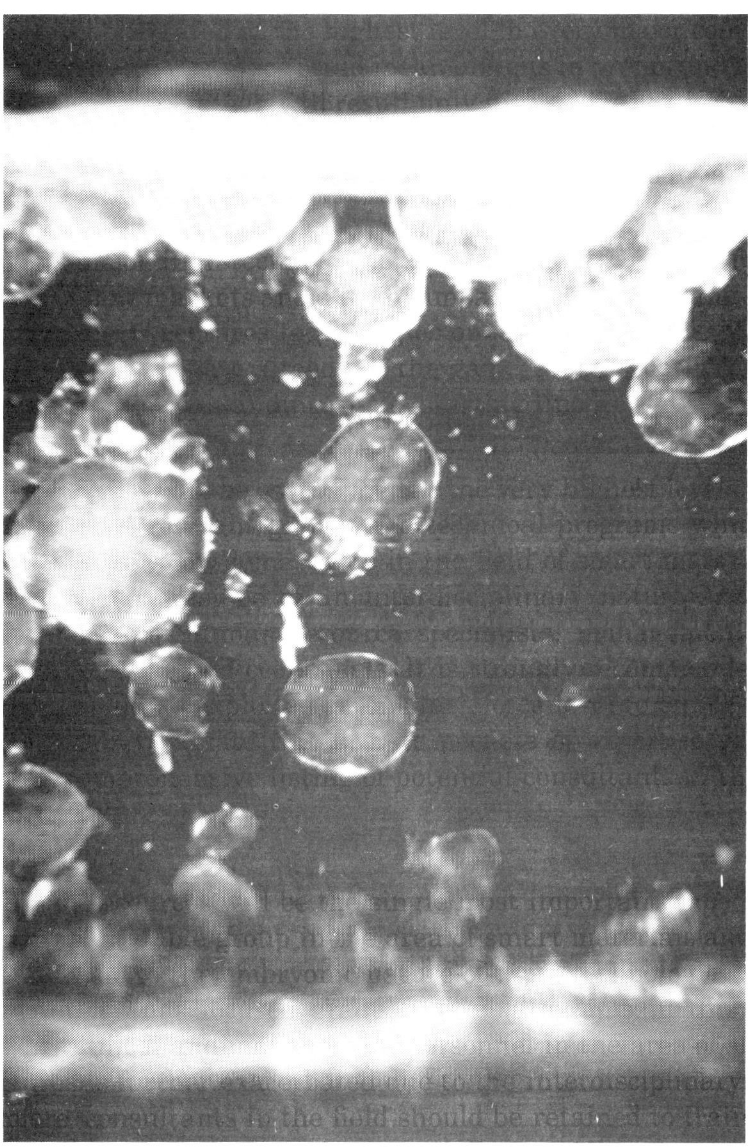

FIGURE 4.2 Photomicrograph of ER fluid with 0 kV/mm field strength.

FIGURE 4.3 Photomicrograph of ER fluid with 2 kV/mm field strength.

Figures 4.2 and 4.3 present photomicrographs of an electro-rheological fluid subjected to electrical field intensities of 0 kV/mm and 2 kV/mm respectively. The photomicrographs were taken in the Intelligent Materials and Structures Laboratory at Michigan State University using a Zeiss Universal phase-contrast microscope with a ×40 magnification and a Chinon camera. The black regions at the top and bottom of the photographs are images of the electrodes employed to generate the electrical field in the ER fluid.

Figure 4.2 clearly shows the random structure of the suspension when a potential difference is not generated between the electrodes. Figure 4.3 clearly shows the truly dramatic change in the structure of the suspension upon developing a potential difference between the electrodes of magnitude 2 kV/mm. Under these conditions, the particles in the suspension orientate themselves in relatively regular chain-like patterns to form a mixture with globally anisotropic mechanical properties. These columnar structures restrict the fluid motion, thereby increasing the energy dissipation or viscous characteristics of the suspension in addition to increasing the stiffness characteristics of the fluid suspension.

Furthermore, a re-orientation of the mass distribution of the suspended particles relative to the electrodes is associated with the change in the mass distribution of the smart structure. Thus, by imposing an electrical field upon an ER fluid, the mass, stiffness and energy-dissipation, or damping, characteristics of the electro-viscous suspension are changed. When the field returns to a zero potential, upon switching off the electrical energy supplied to the electrodes, the particles return to a state of random orientation in the carrier fluid as shown in Figure 4.2.

The voltages required to activate the phase-change in ER fluids are typically in the order of 4 kV per millimeter of fluid thickness, but since current densities are in the order 10 $\mu A/cm^2$, the total power required to trigger this phenomenon is quite low. Furthermore the response of these ER fluids to the excitation voltage is typically less than 1 millisecond, since the fluids can respond to voltage pulses with frequencies up to approximately 11 kHz, which is far beyond the requirements of most practical devices. Electro-rheological fluid suspensions typically have viscosities of the order of 50cP in the uncharged state, and some fluids are able to withstand a non-vanishing yield stress up to 60kPa and shear rates up to 400/s. When the material is stressed below this threshold, it behaves as a solid, but as the stresses are increased, then fluid flow occurs but the yield stress remains nominally constant as shown in Figure 4.4.

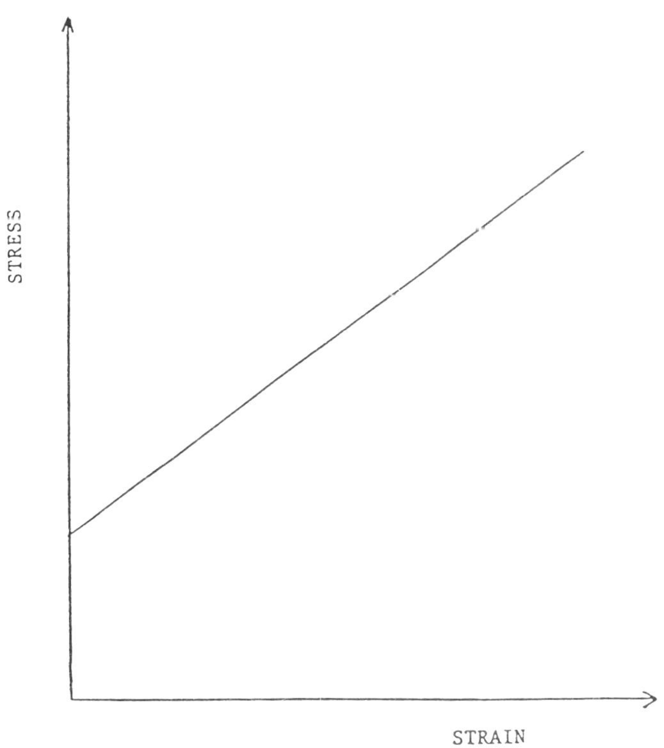

FIGURE 4.4 Rheological characteristics of a typical electro-rheological (ER) fluid.

ER Fluid Technologies

Electro-rheological (ER) fluids have a significant role to play in the impending revolution in smart materials technologies. The specific application naturally dictates the specific characteristics of the particular class of fluids to be employed in the proposed smart system. Field-effect fluids were first reported in the literature in 1880 in Germany, but the field lay dormant until 1947 when Winslow filed a U.S. patent for a "wet" fluid (a "Winslow" fluid). Significant strides in ER fluids technology were made by Stangroom's group in Sheffield, England in the late 1970s. Other significant accomplishments that have made ER fluids commercially viable have been made by Block at Cranfield Institute of Technology in the late 1980s where he developed a substantially waterless fluid. Further enhancements in the technology have been made by Filisko and Armstrong at the University of Michigan who patented a waterless fluid in 1988. These significant milestones in the evolution in ER technologies are highlighted in the subsequent sections.

Winslow's Electroviscous Fluids (WET)

Winslow's fluids employed finely divided solids such as starch and gelatin that were suspended in a nonconductivity liquid such as transformer oil or mineral oil. The particles were dispersed by employing various dispersants. Later silica gels and silicone oils were employed to develop a new class of ER fluids.

Limitations

- abrasive fluids
- high power consumption
- particles settle out of suspension
- these water-based fluids ("wet" fluids) have a restricted service envelope temperature range ($-20°C$ to $70°C$)

Stangroom's Electro-Rheological Fluids (WET)

Stangroom's fluids (1977) were substantially superior to the previous generation of ER fluids because of their reduced abrasiveness, and they were less susceptible to the settling of the particulate phase during prolonged periods of inactivity. These fluids employed polymers such as polyhydric alcohols dissolved in an oleaginous medium. A necessary condition for this fluid was that the alcohol component was almost insoluble in water and the oleaginous medium. Later Stangroom employed a wide variety of polymers comprising acid groups to develop ER fluids with electroviscous effects at least several times superior to the previous generation of fluids. Since Stangroom's fluids contain substantially lower amounts of water, the power requirements are much smaller. Stangroom and Harness later developed fluids employing cross-linked polymers and ethers as the liquid medium that have extended the operating temperature range and reduced the power consumption. Moreover they operate over a wide range of field strengths. Several desired characteristics are highlighted below:

Advantages

- extended temperature range
- reduced power consumption
- broad range of field strengths
- superior chemical and mechanical stability
- slight potential for commercialization
- low viscosity at zero field strength

Block's Fluids (NEAR-DRY)

Block's fluids typically contain less than five percent water by volume. These fluids comprise particles of semi-conductor materials, which conduct electricity. Typically, organic service conductors, transition metals and polymeric semiconductors have been employed in this class of near-dry fluids.

Advantages

- broad temperature range −40°C to 150°C
- low viscosity at zero field strength
- low power consumption
- high dielectric strength
- wide range of electrical field strengths
- minimal settling of the particulate phase

Disadvantages

- thermal problems

Filisko and Armstrong's Field-Effect Fluids (DRY)

The University of Michigan fluid features a non-conducting liquid in which particles of crystalline zeolite (aluminum silicates) are suspended. These particles have very large surface areas due to their "sieve-like" configurations.

Advantages

- completely "dry" fluids
- operating range up to 250°C
- good suspension stability
- fluids perform better at higher temperatures

Discrete Applications: Devices

The traditional applications for ER fluids have been hydraulic devices and discrete systems. Typical examples of these applications are discussed below prior to contrasting them with the new generation of smart continuum structural applications featuring these fluids.

Clutches

Clutches featuring ER fluids permit the design engineer to develop a device in which the voltage applied to the fluid domains controls the slippage between the two shaft systems as shown in Figure 4.5.

For zero electric field strength, the input shaft system rotates independently of the output shaft system. In this configuration, a fluid with very low viscosity is preferable. Since the apparent viscosity is a function of the applied voltage, this electrical input permits the input shaft to drive the output shaft through a state of controlled slippage.

Valves

A typical electro-rheological-fluid-based valve arrangement for controlling the flow of a fluid is schematically shown in Figure 4.6. This class of valves will have a tremendous impact on the evolution of a new generation of hydraulic systems with very few moving parts.

Vibration Damping Applications: Spring-Mass Systems

Discrete damping applications, as shown in Figure 4.7, are numerous and generally comprise a spring-mass system that is employed in conjunction with an ER-

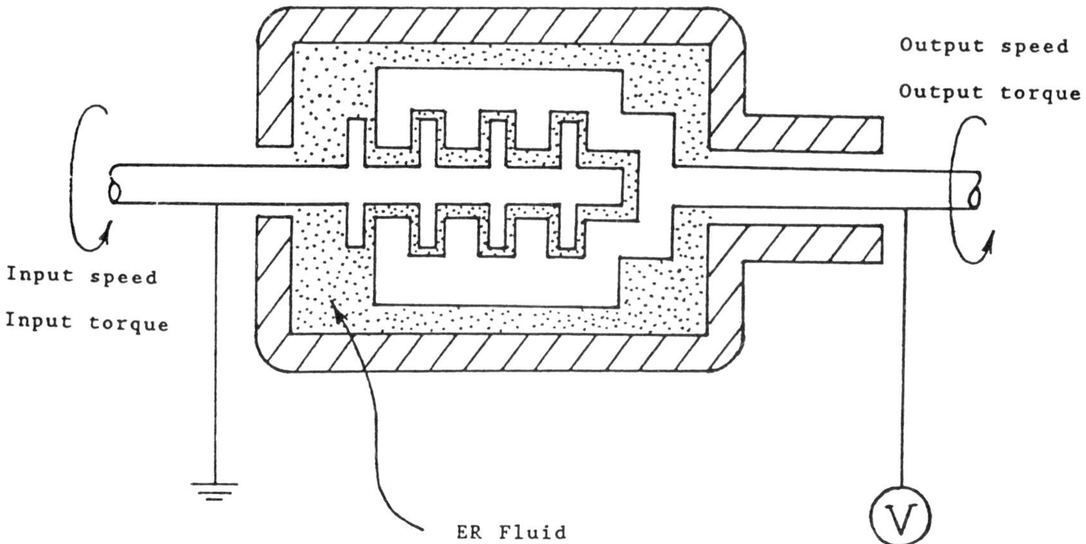

FIGURE 4.5 ER-based clutch.

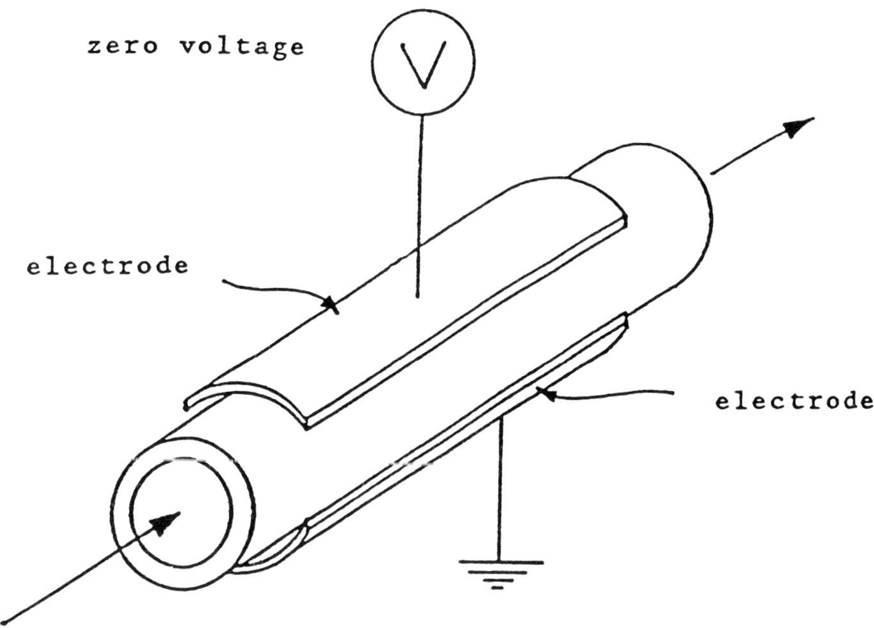

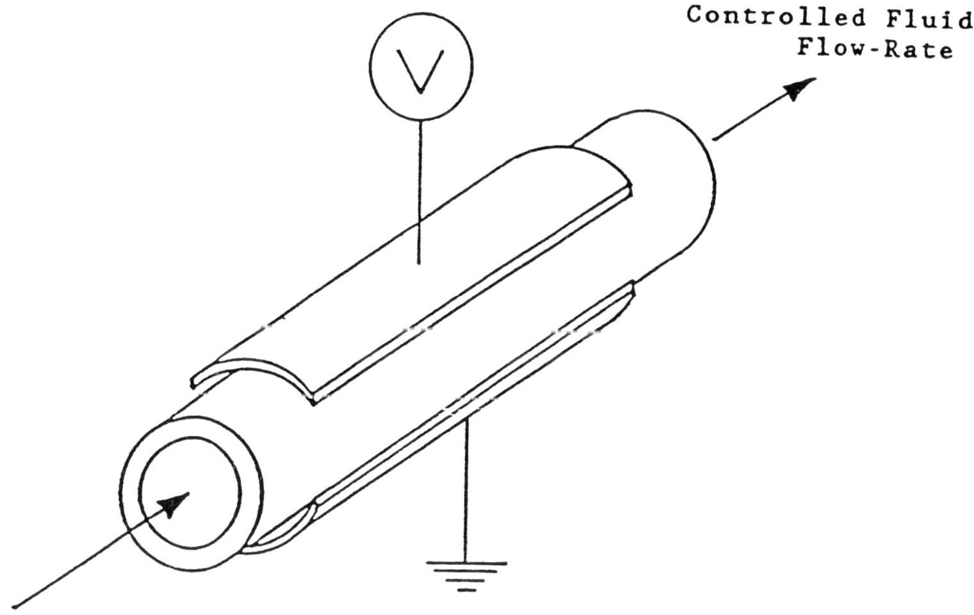

FIGURE 4.6 ER-based valve.

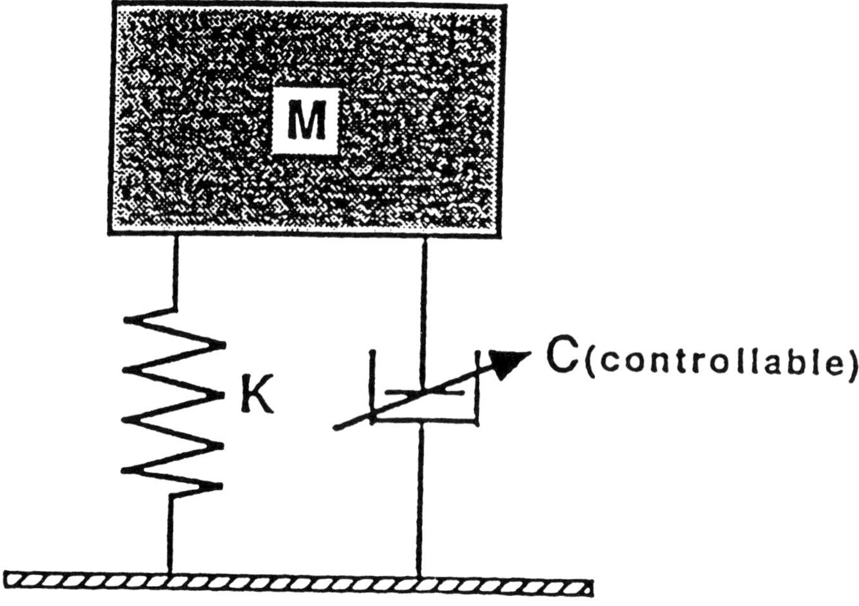

FIGURE 4.7 Vibration of an electro-rheological (ER)-based discrete damper.

based damper. When the overall system is subjected to transient or forced external vibrations by varying the electrical field, the apparent viscosity of the ER fluid in the damper, and hence the system behavior, can be varied continuously in order to comply with system performance specifications. Several orders of magnitude reduction in vibration and levels of acoustical radiation have been achieved by these ER-based devices.

Hydraulic Devices

Traditional hydraulic systems have various problems, such as leakage and seal friction, however, ER fluid devices can be designed in a creative manner in order to overcome these undesirable characteristics.

Foundations for Reciprocating Machinery

Hydraulic and inertia mounts have been traditionally employed to isolate the undesirable vibrational effects of machinery from their foundation. ER-based mounts can be designed in order to tailor the dynamic stiffness to particular vibrational frequencies thereby minimizing vibration transmissibility and noise.

Robots

A typical electro-rheological-fluid-based robot is schematically shown in Figure 4.8. This robot features a smart forearm incorporating embedded ER-domains for vibration control and ER-based joint actuators. This class of robots will yield higher accuracy, shorter settling times and faster response times.

Fixturing Applications

A smart adaptable fixture featuring an array of ER-based "plug-in" removable modules is presented in Figure 4.9. This concept generalized the traditional "bed-of-nails" flexible fixture concept and imparts smartness by virtue of controllable piston characteristics of the removable module. Smart fixtures and grippers of this class address a critical bottleneck highlighted by the U.S. Congress Office of Technology Assessment (OTA), and are crucial for the evolution of the factory of the future.

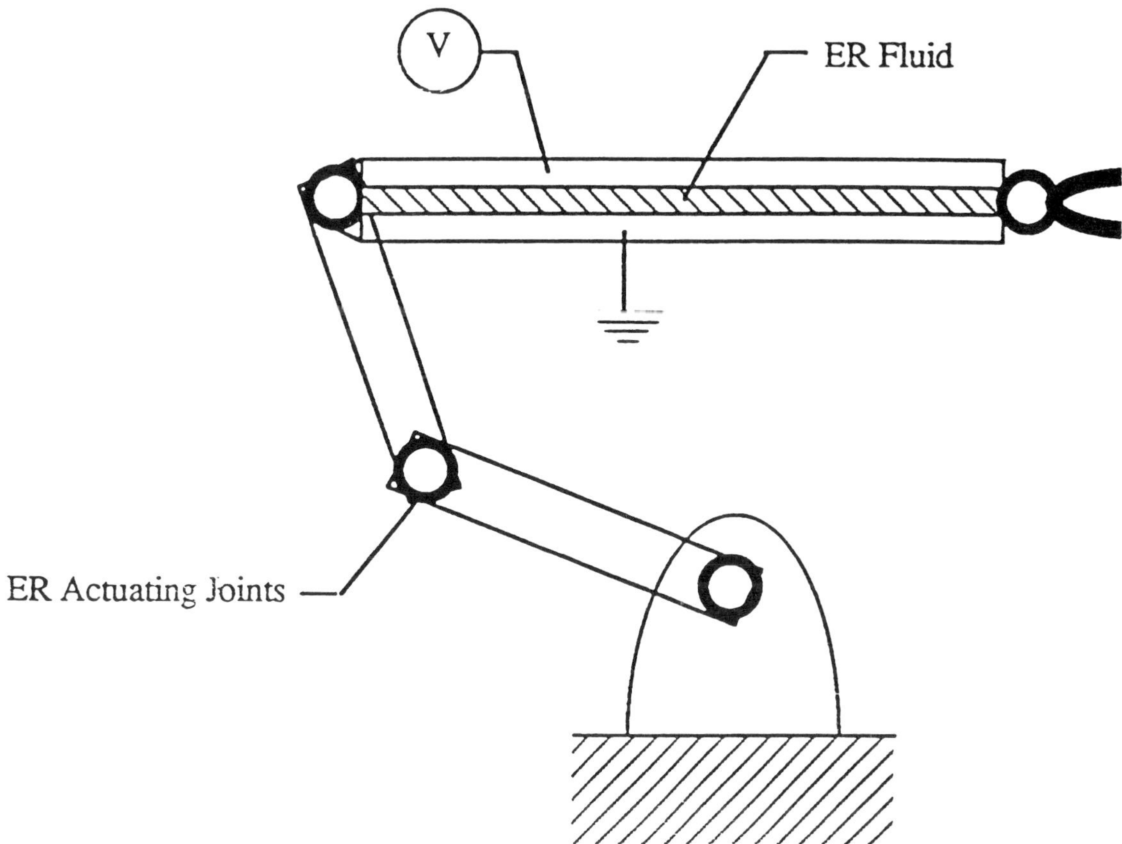

FIGURE 4.8 Electro-rheological (ER)-based robot.

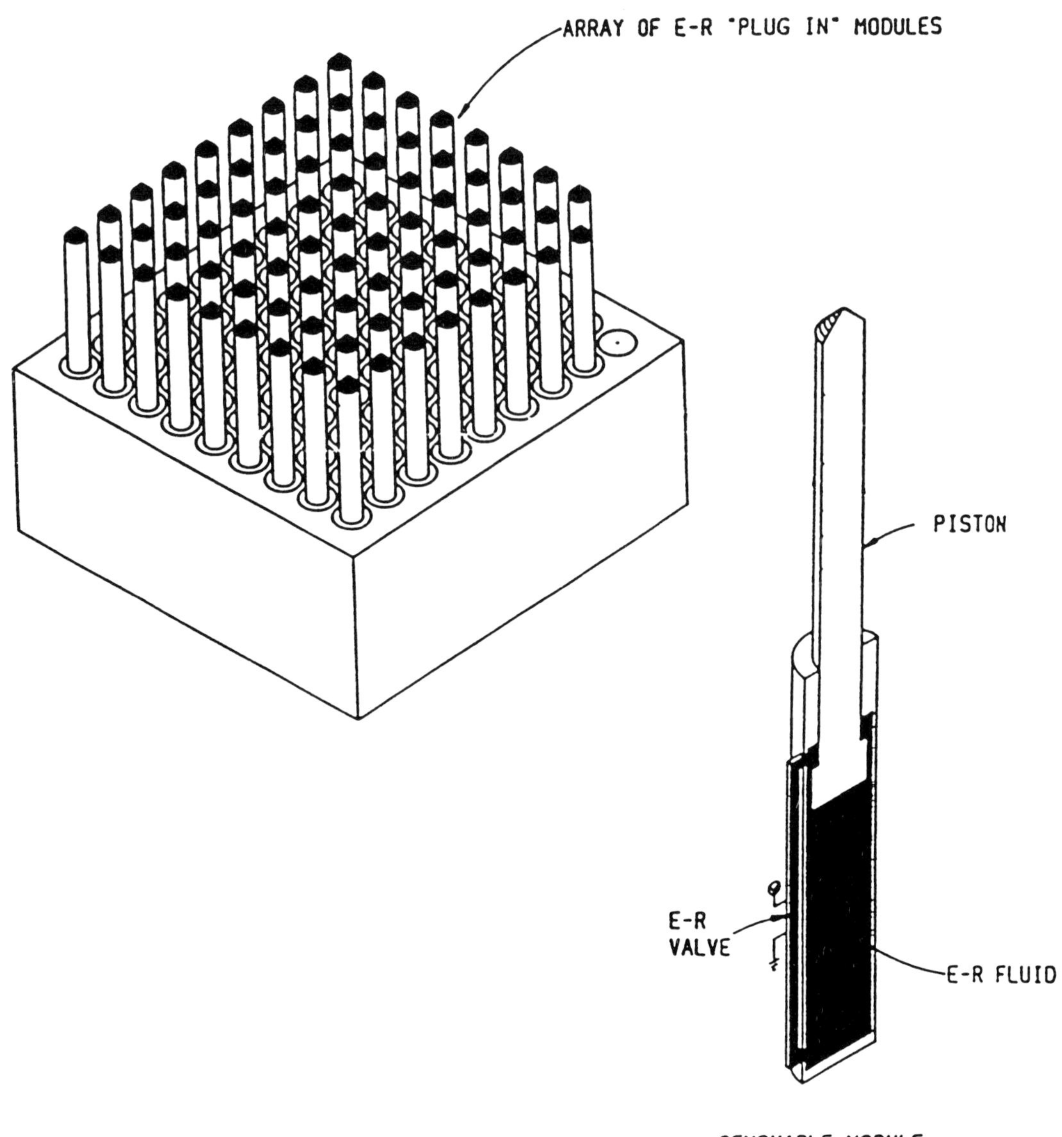

FIGURE 4.9 Electro-rheological (ER)-based adaptable fixture.

Continuum Applications: Structures and Machine Systems

A new generation of revolutionary, intelligent, ultra-advanced composite materials featuring electro-rheological fluids for the active continuum vibrational control of structural and mechanical systems has been developed at the Intelligent Materials and Structures Laboratory at Michigan State University. These ultra-advanced composite materials capitalize on the superior characteristics of advanced composite materials that are interfaced with dynamically-tunable ER fluids contained in voids in the advanced composite structure as shown in Figure 4.10. A variety of geometric configurations of ER fluid domains embedded in structural materials may be employed to optimize the vibrational characteristics of smart structures as shown in Figure 4.11. Alternatively, a laminated configuration may be employed as shown in Figure 4.12.

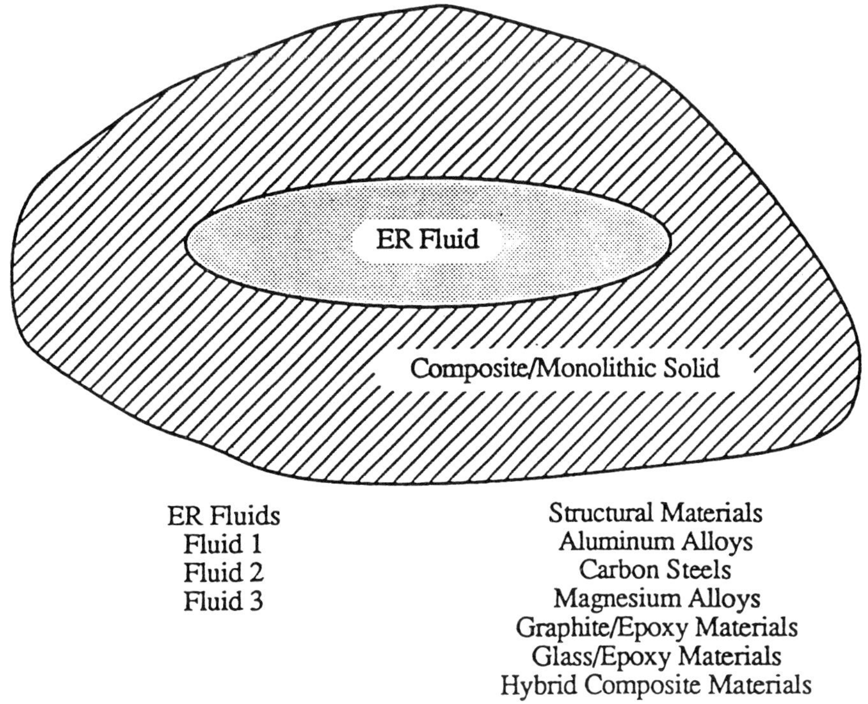

FIGURE 4.10 Schematic configuration of ultra-advanced smart materials incorporating electro-rheological (ER) fluids.

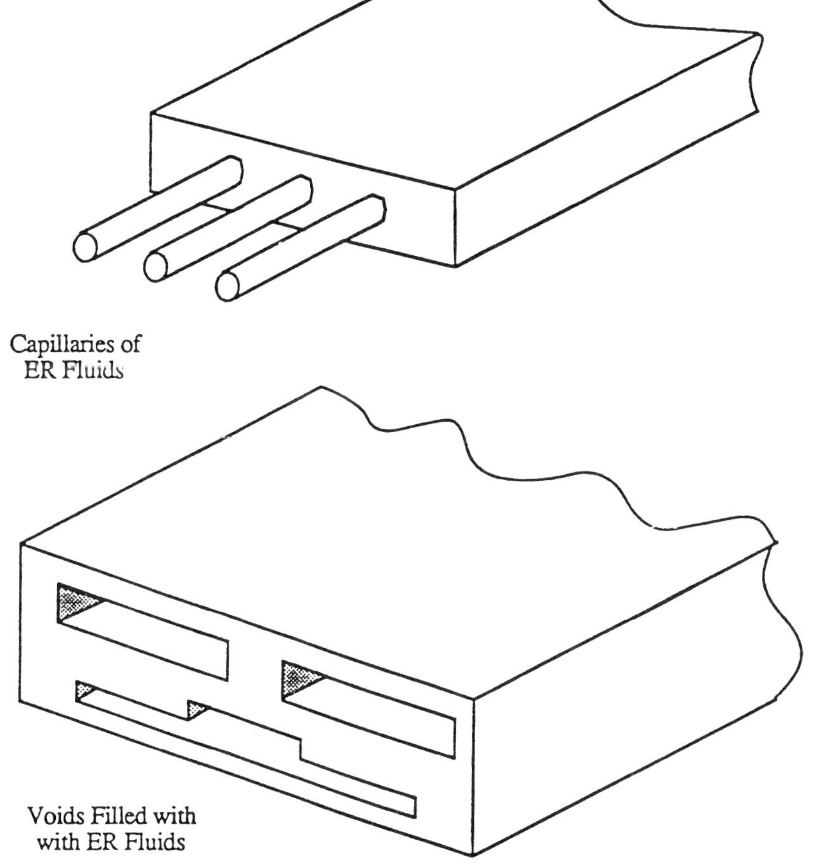

FIGURE 4.11 A variety of geometrical configurations of ER fluid domains embedded in structural materials to optimize the vibrational response characteristics of smart structures.

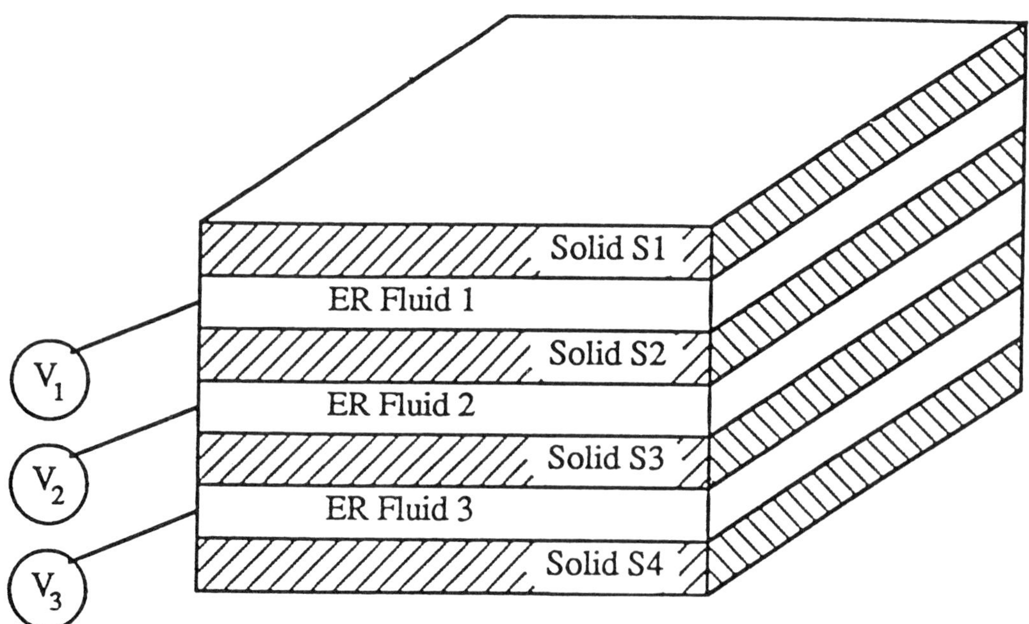

FIGURE 4.12 Laminated electro-rheological (ER)-based composite structure.

Changes in the electrical field imposed upon the ER fluids dramatically alter the rheological characteristics of the fluids and hence the global mass, stiffness and dissipative characteristics of the ultra-advanced composite structure. The instantaneous response-time of the ER fluids and the inherent ability of these materials to interface with solid-state electronics and modern control systems provides designers with a unique capability to synthesize ultra-advanced intelligent composite structures whose continuum electro-elastodynamic response can be actively controlled in real-time. An application of this philosophy to control the vibrational response of a smart aircraft wing is schematically represented in Figure 4.13. Typically, a wing of this kind would feature smart plate and shell-like substructures as shown in Figure 4.14.

This class of innovative materials derives its intelligence from the merger of sensors, built into the finite element control segments of the ultra-advanced composite material continuum, microprocessors, and dynamically-tunable electro-rheological fluids as shown in Figure 4.13. The sensors monitor the elastodynamic behavior of the ultra-advanced composite structure, and the signals from the sensors are fed to the appropriate microprocessor that evaluates the signals prior to determining an appropriate control strategy in order to synthesize the desired elastodynamic response characteristics. This synthesis is typically accomplished by controlling the rheological characteristics of the ER fluid domains in the finite element segment associated with the particular sensor. Alternatively, intelligent sensors may be employed by integrating the sensing function and the associated electronic data processing function on a single I.C. chip. This change in the rheological characteristics of the ER fluids in a typical finite element control segment, in turn, alters the global mass, stiffness, and damping characteristics of the ultra-advanced composite structure in order to achieve the desired vibrational response.

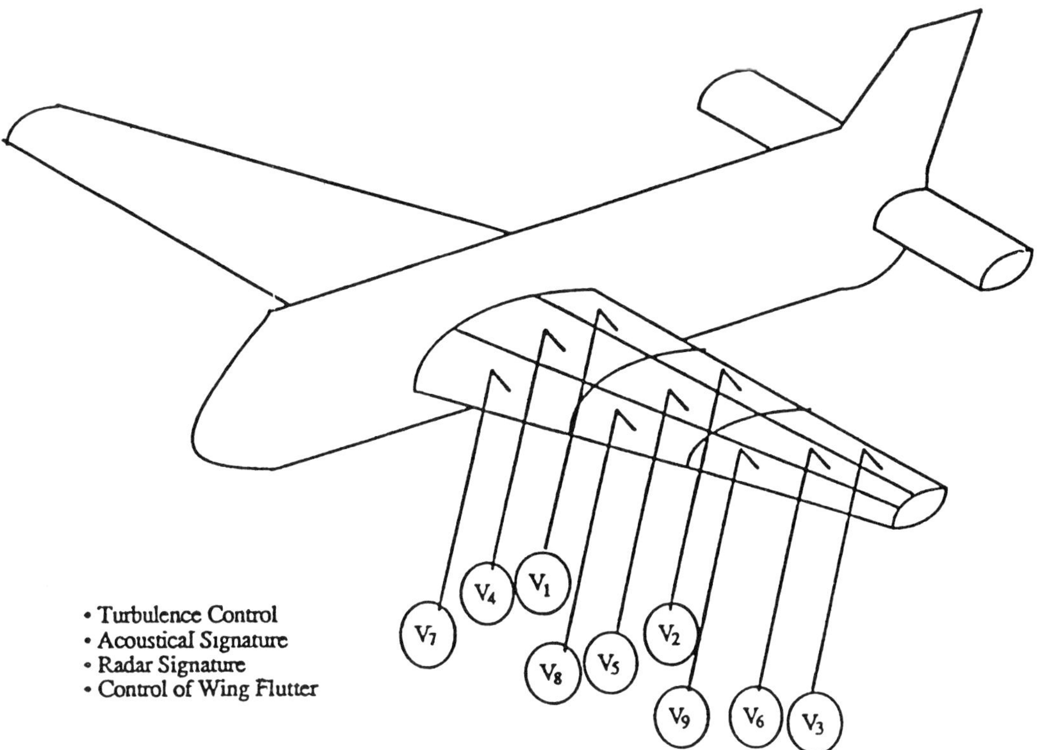

FIGURE 4.13 Smart aircraft wing.

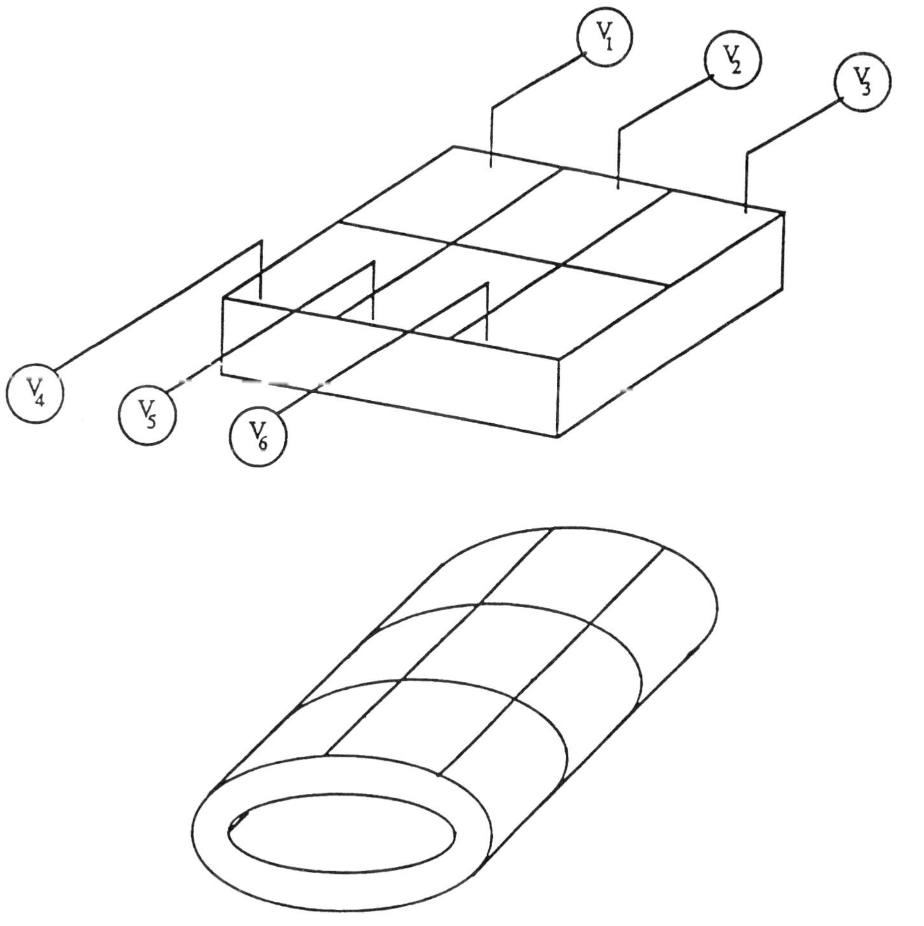

FIGURE 4.14 Shell-like structure with embedded electro-rheological (ER) fluid domains.

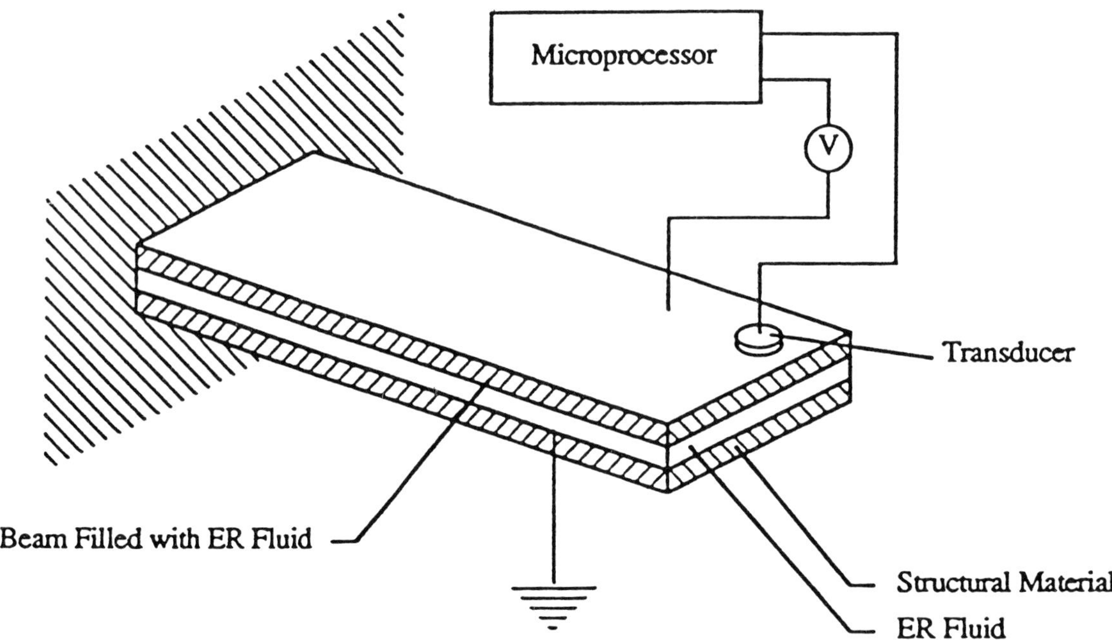

FIGURE 4.15a Ability of smart electro-rheological (ER)-based beams to dramatically change their vibration characteristics: schematic of the beam.

Pioneering proof-of-concept experimental investigations focused on evaluating the elastodynamic response of cantilevered beams fabricated from ultra-advanced intelligent composite materials have been undertaken at the Intelligent Materials and Structures Laboratory at Michigan State University (IMSL/MSU). The preliminary results of this investigation clearly demonstrate for the first time the ability to dramatically change the transient elastodynamic response characteristics of beam-like specimens fabricated in ultra-advanced composite materials by changing the electrical field imposed on the ER fluid domains, as shown in Figures 4.15a and 4.15b.

These proof-of-concept investigations highlight the tremendous potential for actively controlling the elastodynamic response of continuum structures and machine systems fabricated in ultra-advanced composite materials featuring ER fluids. This potential can be exploited by extending the fundamental phenomenological work synoptically presented on the previous pages by the successful incorporation of intelligent sensor technologies and modern control strategies. Various pioneering proof-of-concept investigations have been undertaken by IMSL/MSU on beam and platelike structures, slider-crank mechanism, robots, and helicopter rotors; the salient results of these investigations are presented in Figures 4.16–4.23. It is anticipated that the successful transition of this embryonic technology into an emerging technology would significantly accelerate the evolution of this innovative class of multi-functional, dynamically-tunable, ultra-advanced, intelligent, composite materials for military, aerospace and advanced manufacturing environments of the future.

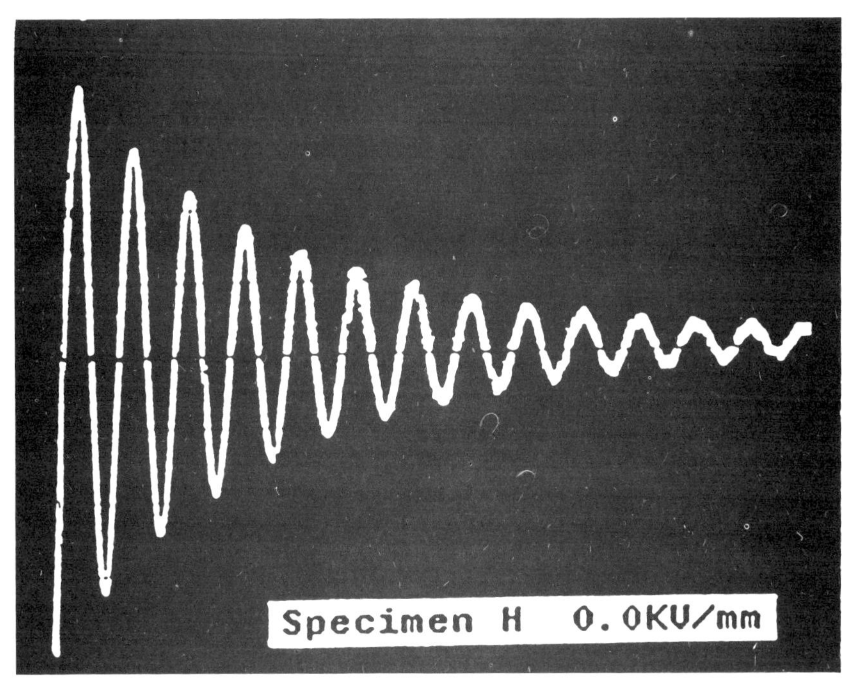

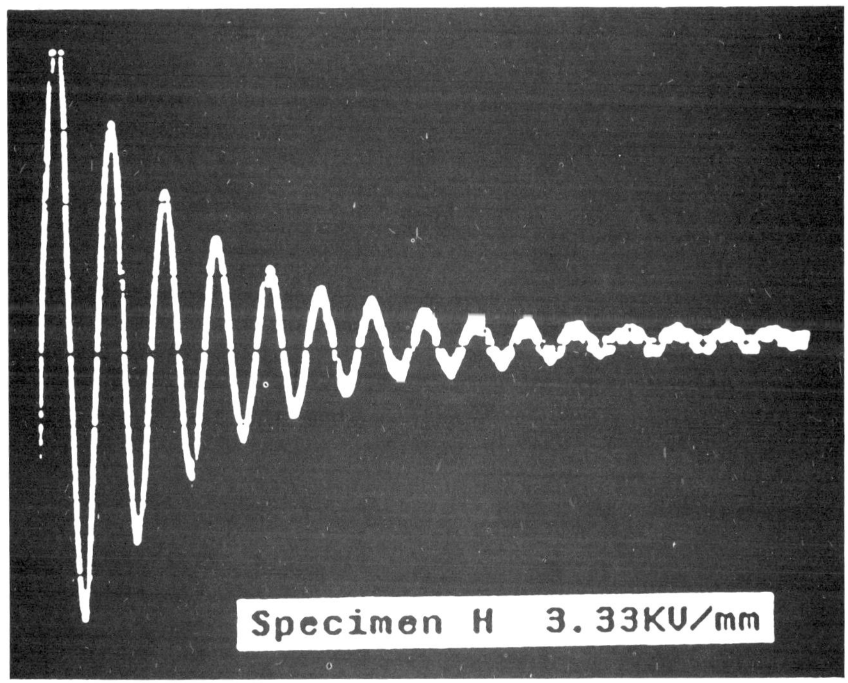

FIGURE 4.15b Ability of smart electro-rheological (ER)-based beams to dramatically change their vibration characteristics: experimental results.

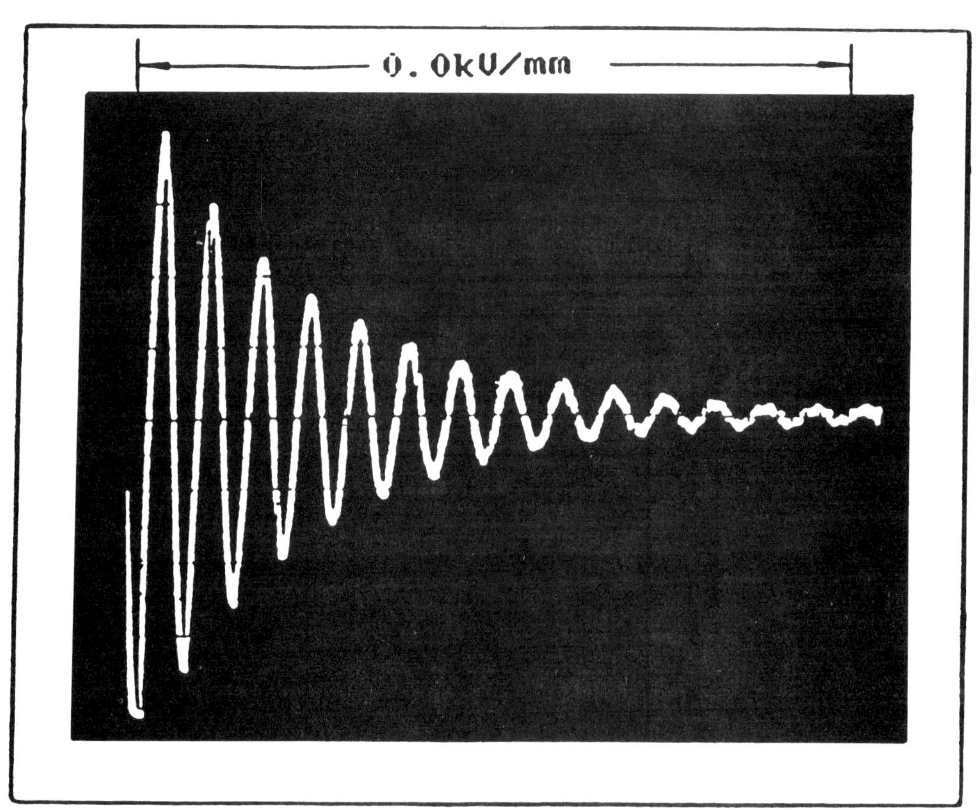

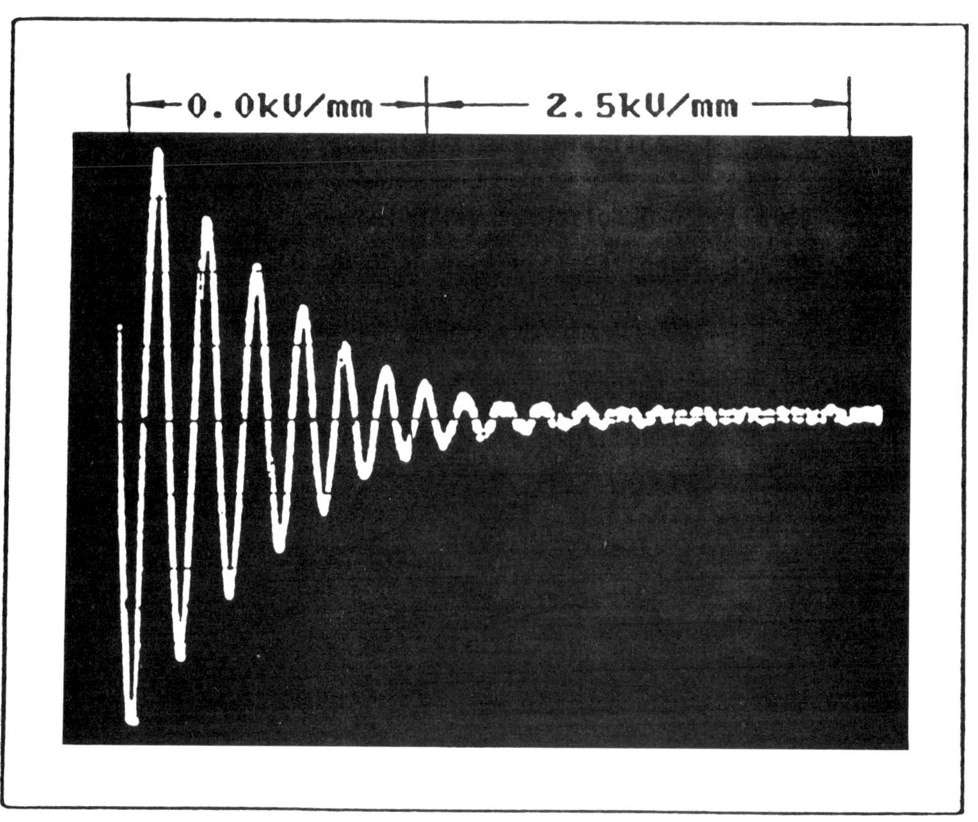

FIGURE 4.16 Transient vibration characteristics of a dynamically-tunable electro-rheological (ER)-based small beam controlled by a bang-bang control strategy: experimental results.

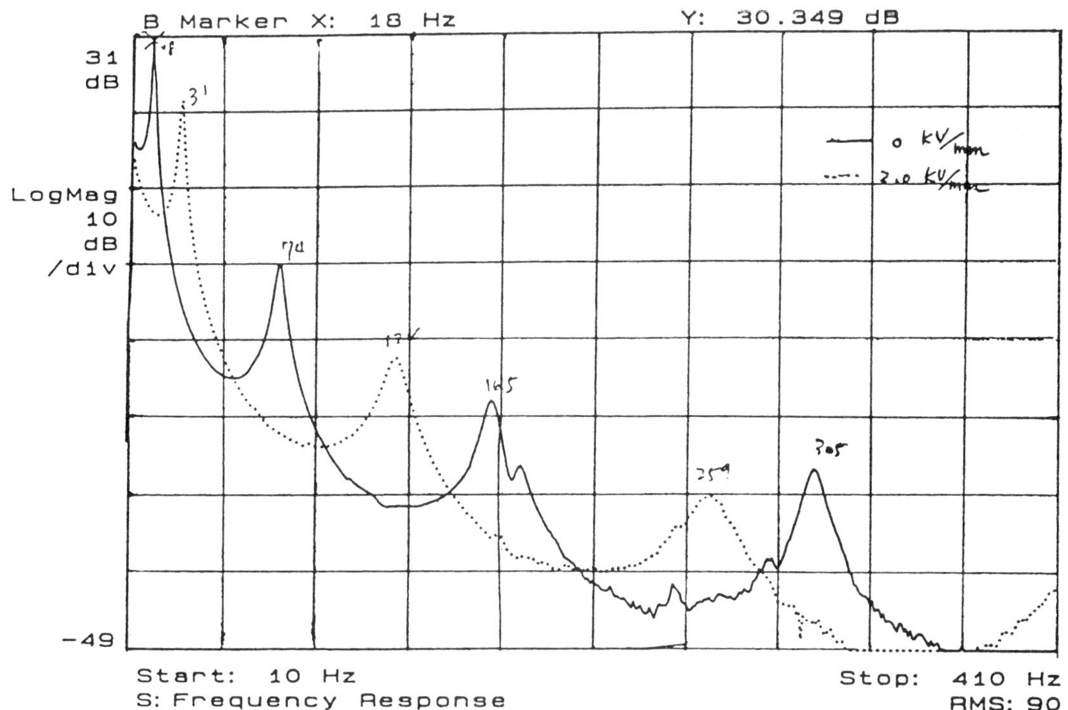

FIGURE 4.17 Unique capability of electro-rheological (ER)-based smart beams to dramatically change their natural frequencies and damping characteristics in real-time.

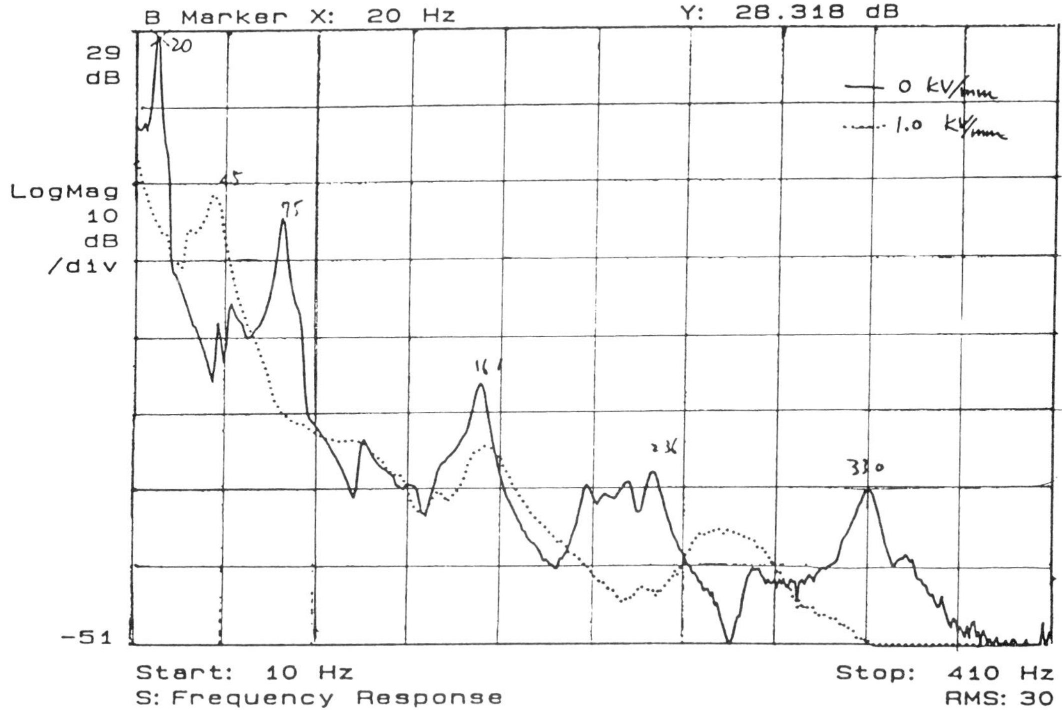

FIGURE 4.18 Unique capability of electro-rheological (ER)-based smart plates to dramatically change their natural frequencies and damping characteristics in real-time.

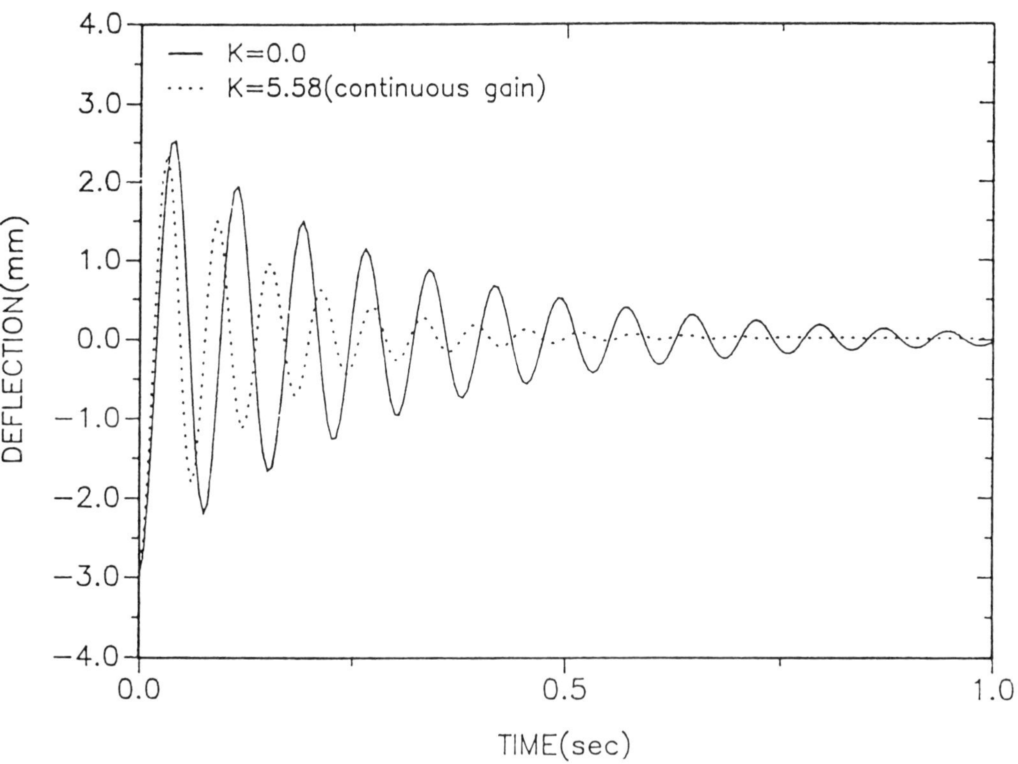

FIGURE 4.19 Development of control strategies for smart electro-rheological (ER) beams to optimize system performance for variable service conditions and unstructured environments.

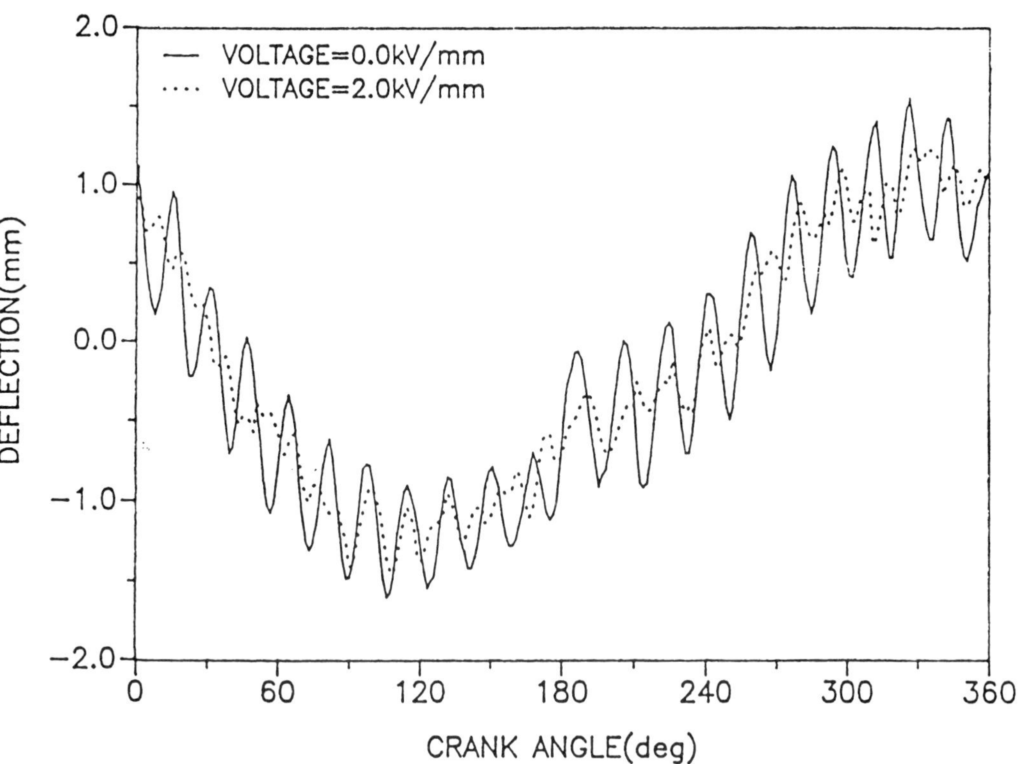

FIGURE 4.20 Experimental results demonstrating the real-time control of a smart slider-crank mechanism featuring embedded electro-rheological (ER) fluid domains.

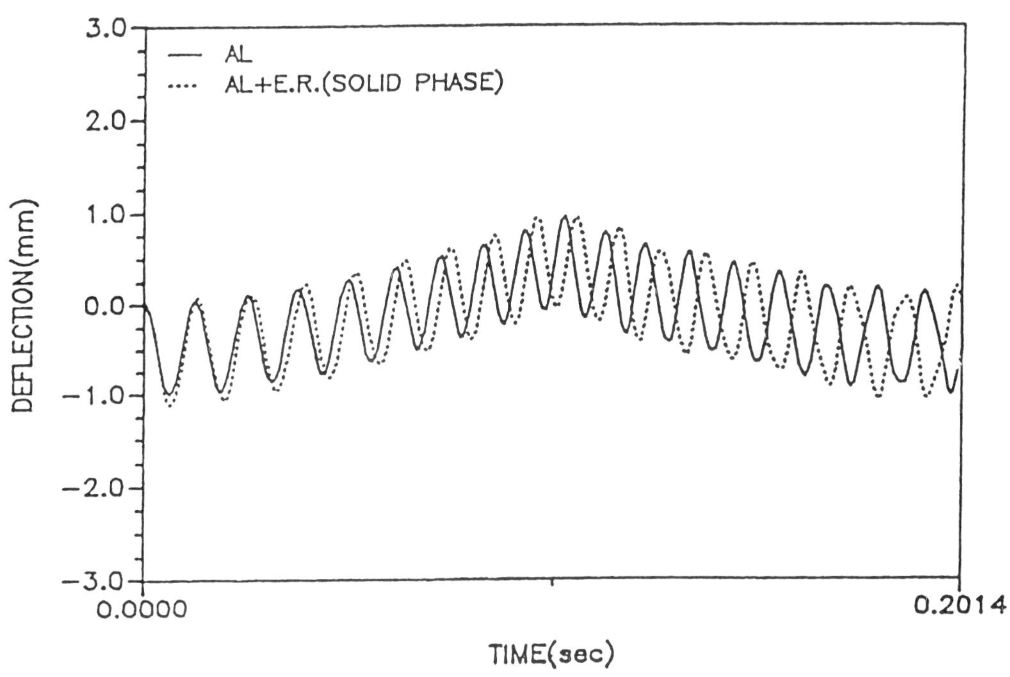

FIGURE 4.21 Computer simulation results demonstrating the capability of a GE-P50 robot retrofitted with a smart arm incorporating embedded electro-rheological fluids.

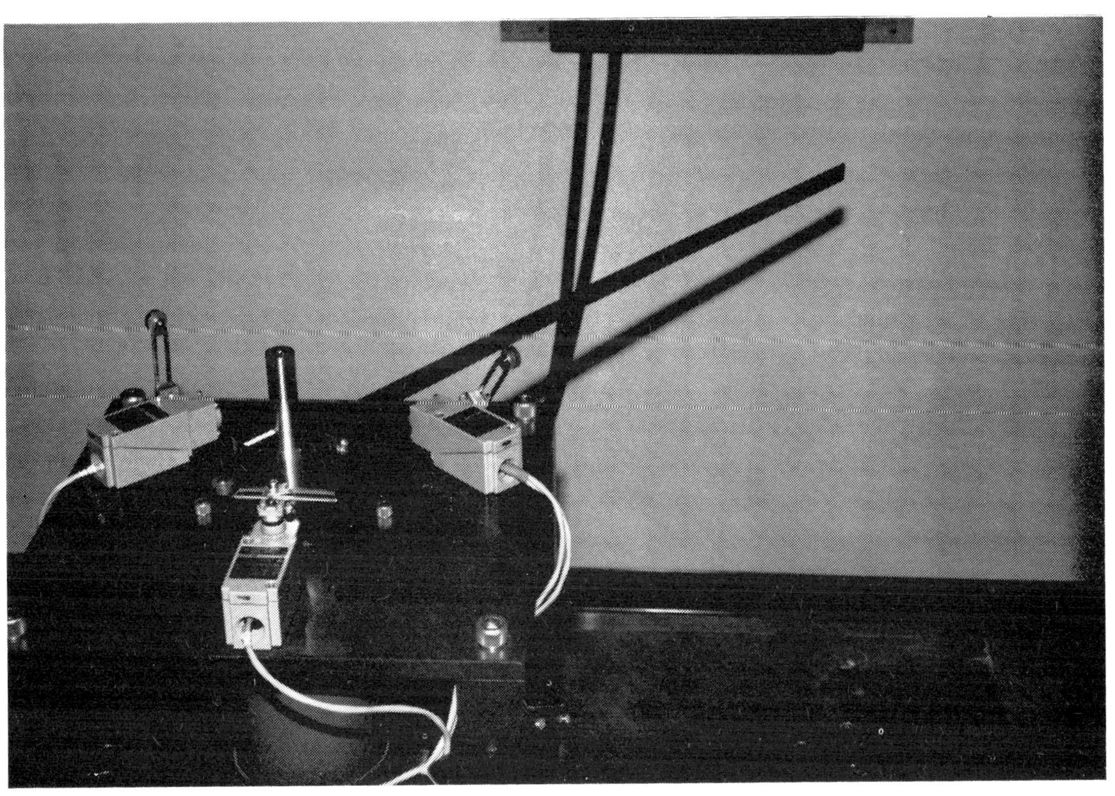

FIGURE 4.22 Proof-of-concept experimental investigation of a smart single-link robot.

FIGURE 4.23 Proof-of-concept experimental investigation on a smart helicopter rotor.

4.2 Shape Memory Alloys

Background on Shape Memory Alloys

The shape memory phenomenon manifests itself when a shape memory alloy (SMA) is plastically deformed in the low temperature martensitic condition and, upon the removal of the external loads, regains its original shape when heated. The exact mechanism by which the shape recovery takes place is not well understood, however, the process of regaining the original shape is known to be associated with a reverse transformation of the deformed martensitic phase to the higher temperature austenitic phase.

Typical materials which exhibit the shape memory effect include the copper alloy systems of Cu-Zn, Cu-Zn-Al, Cu-Zn-Ga, Cu-Zn-Sn, Cu-Zn-Si, Cu-Zn-Ni, Cu-Au-Zn, Cu-Sn, and the alloys of Au-Cd, Ni-Al, Fe-Pt. Nitinol, a nickel-titanium alloy, is the most common of the shape memory alloys or transformation metals.

Nickel-titanium alloys (Nitinol, NiTi) acquire their name from Ni (Nickel)-Ti(Titanium)-NOL (Naval Ordnance Laboratory). NiTi alloys, featuring a near-equiatomic composition, can be plastically deformed in their low temperature martensitic phase and then restored to the original shape by heating them above the characteristic transition temperature. Typically plastic strains as high as six to eight percent may be completely recovered by heating the Nitinol in order to transform it to the austenitic phase; furthermore, constraining it from regaining the memory shape can result in stresses of 100,000 psi, for example. The yield strength of martensitic Nitinol is approximately 12,000 psi. Typical material characteristics for Nitinol are presented in Figure 4.24.

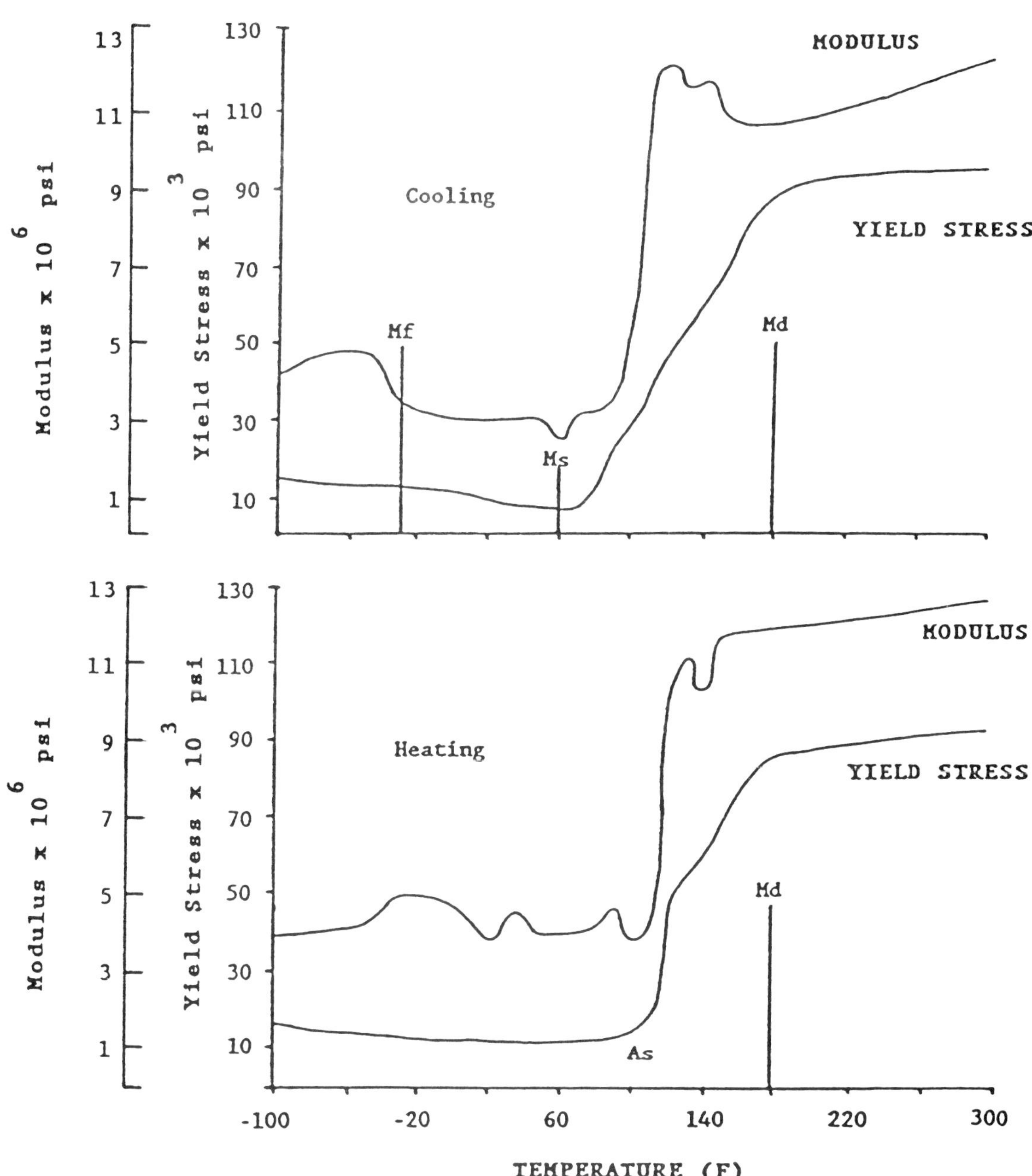

FIGURE 4.24 Typical material characteristics for Nitinol.

There is a substantial database on the thermal, electrical, magnetic, and mechanical characteristics of Nitinol; however, the influence of residual stresses and high temperatures associated with the fabrication and processing of Nitinol-based composites is not well understood. Furthermore, the extent, duration and repeatability of the shape memory effect, as well as the dynamic actuator and sensing characteristics of Nitinol, are also not understood.

Continuum Applications: Structures and Machine Systems

Shape memory alloy-based smart structures are typically fabricated with reinforced composite materials. Fibers of SMA are embedded into either a matrix material, such as an epoxy resin system, or in a laminated fiber-reinforced composite material as shown in Figure 4.25. By changing the temperature beyond the phase-transition point, the shape and mechanical characteristics of the SMA and, hence, the shape and global mechanical characteristics of the smart structure, are changed.

In typical beam, plate and shell-like structures, the SMA fibers or films are embedded in symmetric pairs off the neutral axis/surface in order to control the geometry of the structure. Prior to the fabrication of the smart structure, the SMA fibers are plastically deformed and are constrained in a configuration that is different from their memory configuration at the end of the curing cycle. Typically, in service, the fibers are electrically heated; once the phase-transition temperature is achieved, the SMA fibers have a tendency to regain their memory configuration. In this process a distributed shear load is generated along the length of the fibers due to the off neutral-axis location that results in the controlled application of a bending moment to the smart structure with the attendant change in configuration. Sleeves can be created within the composite laminates in order to accommodate the plastically-elongated shape memory alloy prior to clamping the SMA to both ends. When the shape memory alloy is heated, the fibers will contract if they are not clamped. When one end of the beam is free, the fibers in a sleeve will exert a concentrated force on the ends of the structure in a direction that is always tangential to the structure at the point where the fibers are clamped to the structure. However, when both ends of the beam are fixed, heating the SMA results in fibers with a significantly increased stiffness and applied tension that will resist any transverse motion.

Steady-state vibration control, which may also be used for structural acoustic control, can be accomplished with SMA reinforced composites by employing "Active Modal Modification." The modal response of a structure can be tuned by heating the SMA fibers to change the stiffness of all or portions of the structure. For example, when Nitinol is heated to cause the material transformation from the martensitic phase to the austenitic phase, the Young's modulus changes by a factor of approximately four as shown in Figure 4.24. The stiffness is increased by a factor of four and the yield strength increases by a factor of ten. These changes in the material properties occur due to a phase transformation; they do not result in any appreciable force and do not need to be initiated by any plastic deformation.

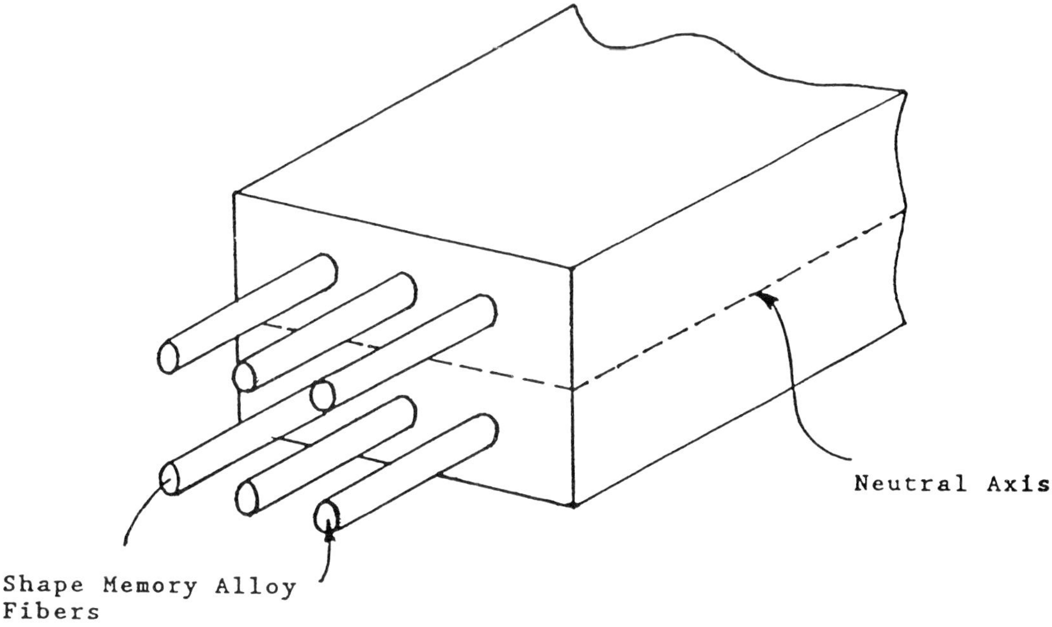

FIGURE 4.25 Smart structures featuring embedded shape memory alloy fibers.

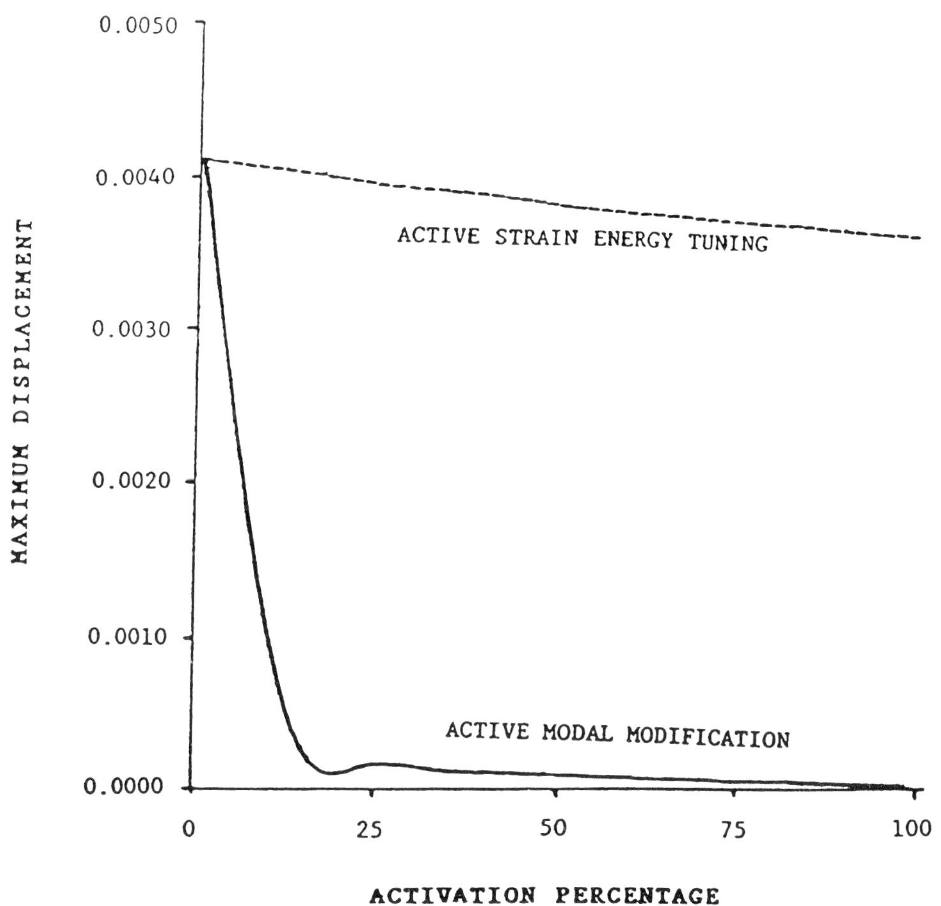

FIGURE 4.26 Maximum plate deflection under a uniform pressure load using active SMA control.

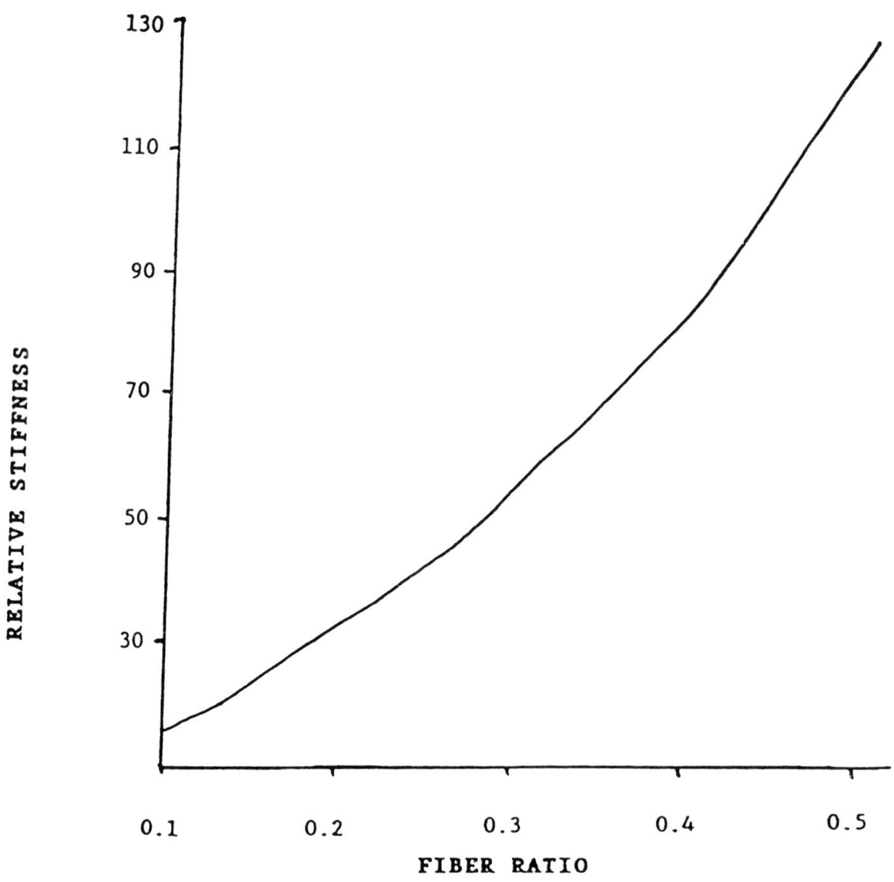

FIGURE 4.27 Flexural stiffness tuning of quasi-isotropic plates using active strain energy tuning.

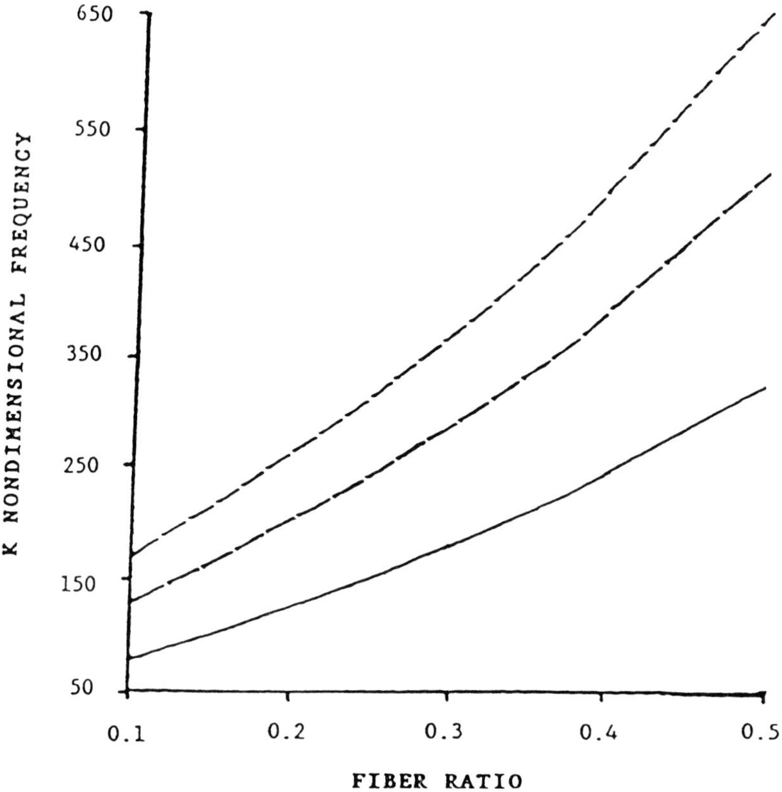

FIGURE 4.28 Variation of natural frequencies as a function of fiber volume fraction.

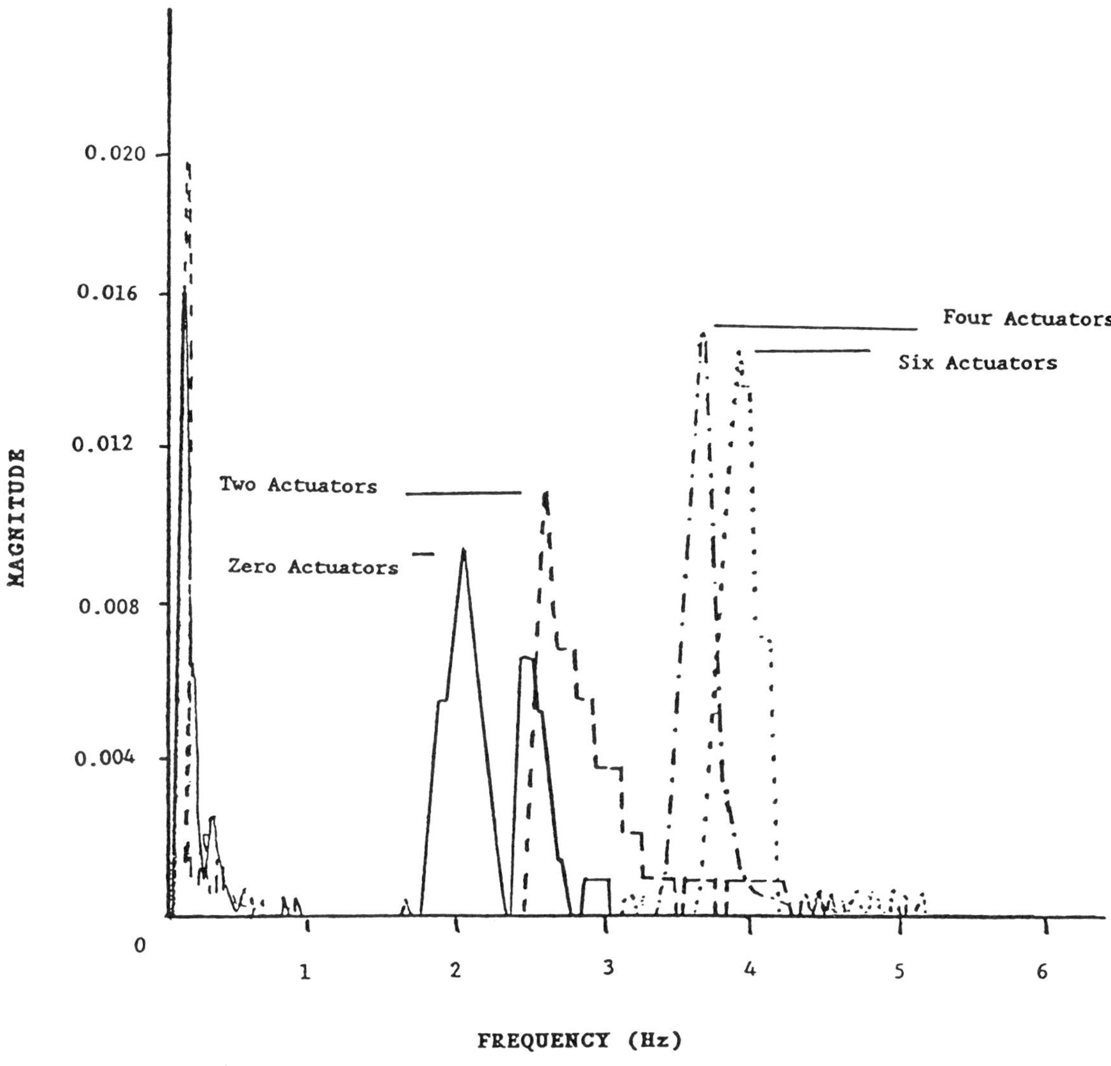

FIGURE 4.29 Natural frequencies of SMA composite beam with zero, two, four, and six actuators.

The shape memory alloy fibers may be placed in or on the structure in such a way that when activated there are no resulting deflections but instead the structure is placed in a residual state of strain. The resulting stored strain energy changes the energy balance of the structure and modifies the elastodynamic response. This response is referred to as Active Strain Energy Tuning.

Active strain energy tuning utilizes both the embedded fiber and sleeve method described earlier. The difference between the embedded fibers and the fibers in a sleeve is that in the first case the force of the shape memory alloy is distributed over the length of the fiber and in the latter case the force is concentrated at the end of the structure, or is used to resist transverse motion. Both of the design concepts described above have been incorporated into prototypes and their potential has been demonstrated on a limited scale. Typical results for plate deflection under a uniform pressure load employing active control of the shape memory actuators is presented in Figure 4.26. Results for the flexural-stiffness tuning of quasi-isotropic plates are shown in Figure 4.27, and the variation of natural fre-

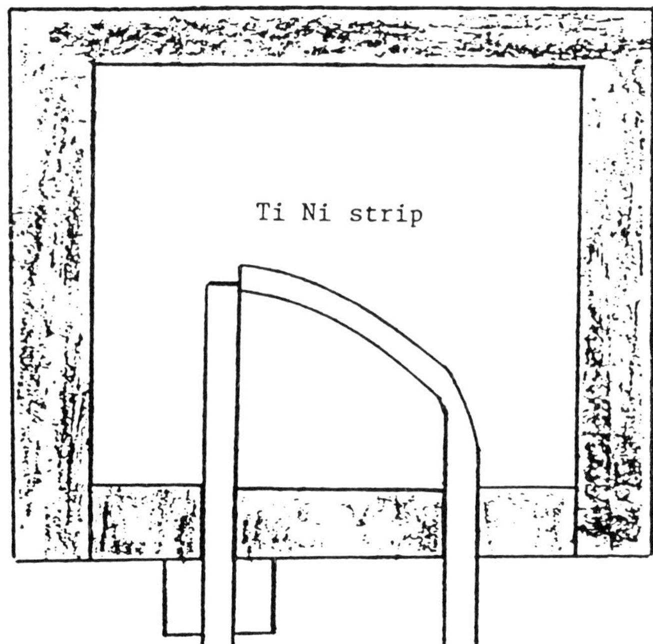

FIGURE 4.30 Temperature fuse employing a shape memory alloy.

quencies as a function of fiber volume fraction and the number of actuators is presented in Figures 4.28 and 4.29 respectively.

Discrete Applications

Historically, SMA based devices have been discrete in nature with typical applications in switching and actuation. A typical application of this kind that involves a temperature fuse is schematically presented in Figure 4.30. However, the Japanese have recently focused their efforts in the field of robotics. Two main types of SMA actuators for robots have been developed, namely, biased and differential. Biasing uses a spring to obtain the bias force against the unidirectional force of the SMA. In the differential type, the spring is replaced with another SMA and the opposing forces control the actuation. Figure 4.31 shows two configurations for SMA actuators. This assembly of a micro-robot actuated by shape memory alloys is schematically shown in Figure 4.32. Recently, researchers at the Intelligent Materials and Structures Laboratory at Michigan State University have developed a new class of machine systems incorporating SMA actuators that have the capability to deliver variable output characteristics as shown in Figure 4.33.

4.3 Piezo-Electric Materials

Piezo-Electric Actuators and Sensors: Background

Piezo-electric materials are materials that generate a charge in response to a mechanical deformation, or provide mechanical strain when an electric field is applied across them. Therefore, piezo-electric materials can be employed either as actuators or sensors in the development of smart structures. Piezo-electric materials are typically crystals, or ceramics; however, due to their brittle characteristics, piezo-electric sensors are typically fabricated in polymeric materials such as polyvinyldene fluoride (PVDF or PVF_2). Thin films of these polymeric materials can be bonded to a surface in much the same way as a conventional strain gauge. Piezo-electric sensors have very high mechanical strength and high sensitivity to changes in pressure. Typically, these sensors are used for tactile sensing, temperature sensing and strain sensing. Typically the polymers have piezo-electric characteristics imparted to them by polarizing them either in a uniaxial or bi-axial film. Uniaxial films may be employed to measure strain in one direction, and bi-axial films measure strain in two directions. Piezo-electric sensors and actuators generate very little heat, they have low power consumption, and since they conserve energy, they are very efficient in several applications as compared with electro-mechanical devices.

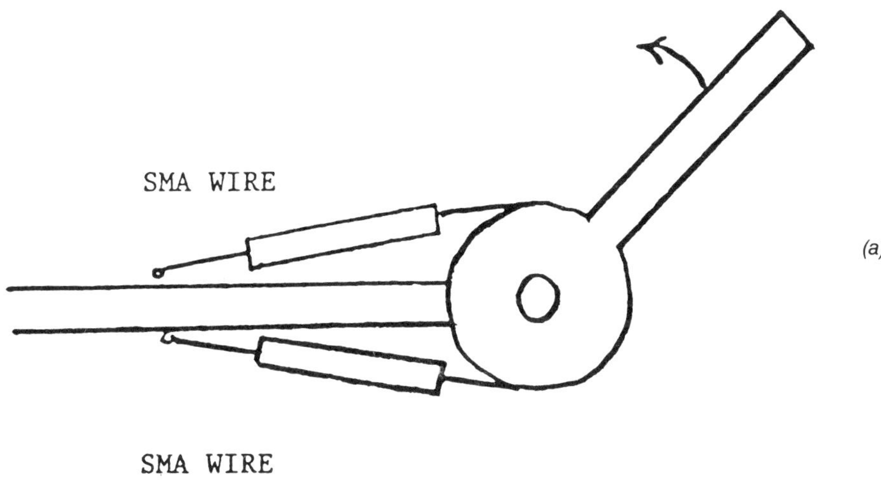

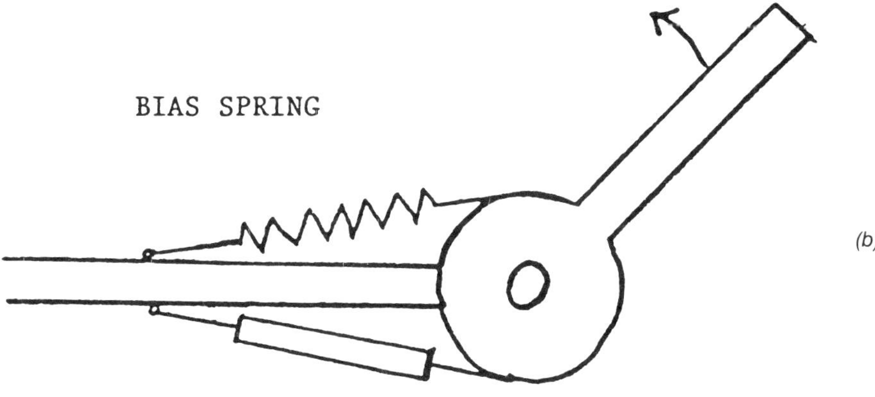

FIGURE 4.31 SMA actuators (a) the bias-type actuator (b) the differential-type actuator.

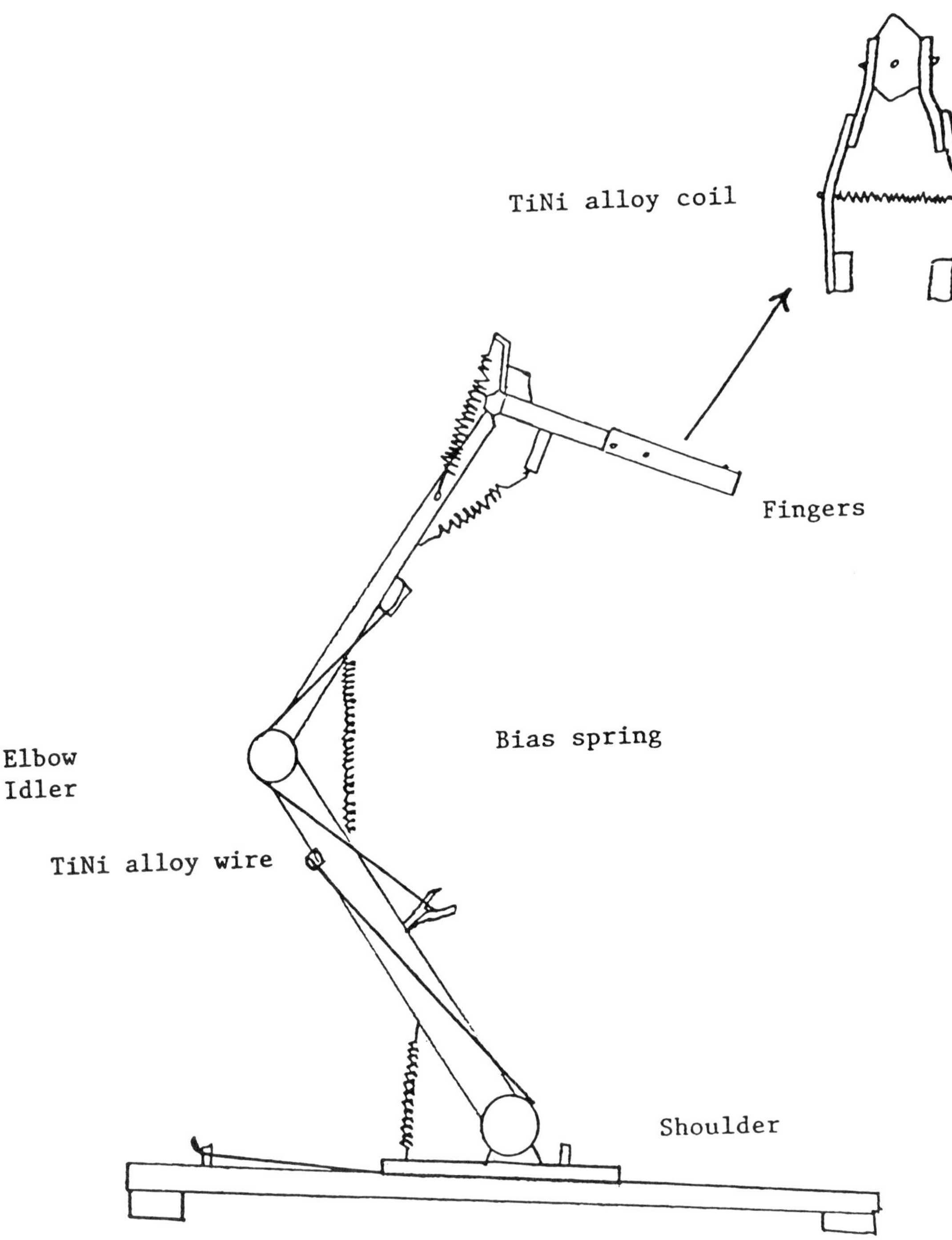

FIGURE 4.32 Assembly of micro-robot actuated by shape memory alloys.

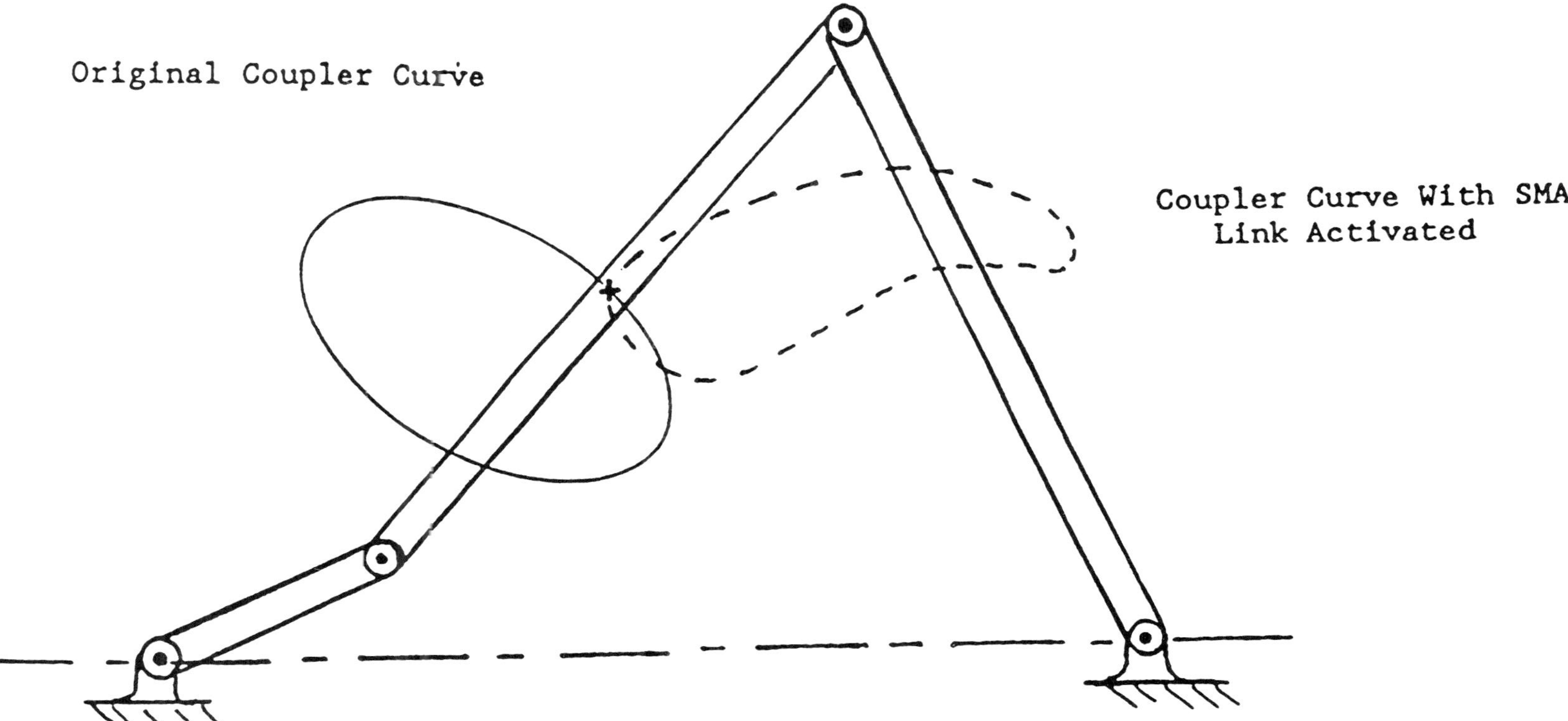

FIGURE 4.33 Four bar linkage with smart shape memory alloy coupler link.

Discrete Applications

Piezo-electric sensors have been employed as tactile sensors to identify objects with a sensitivity sufficient enough to identify the Braille alphabet and various grades of sandpaper. They have also been used as skin-like sensors to mimic the pressure and temperature sensing capabilities of human skin as shown in the University of Pisa robot finger shown in Figure 4.34. Furthermore, they have been employed in high-accuracy situations involving the positioning of printer heads and positioning turntables. They have also been used in finite-degree-of-freedom active vibration damping applications.

Many of the premier accelerometer manufacturers employ piezo-electric crystals in the heart of their transducers. Other discrete applications for piezo-electric devices include the vibration control of turbo-machines.

Continuum Applications

A large number of piezo-electric actuators may be distributed over a structure as shown in Figure 4.35 in order to control the continuum vibrational response of the structure. These lightweight actuators add very little additional mass to the structure thereby preserving the passive mechanical and elastodynamical characteristics of the smart structures. Bonding or embedding finite element segments of piezo-electric actuators would permit the application of controlled localized forces and torques to be imposed on the structure in order to control the vibrational response of the structure. The small-sized, light and simple piezo-electric actuators are ideal candidates for aerospace applications. These smart structures have been developed using aluminum beams, glass-epoxy beams, and graphite-epoxy beams. Whereas the ultimate strength of the laminates was reduced by 20 percent, the global elastic modulus of the beams was not significantly affected.

Piezo-electric actuators with low gain have been employed for the active damping of beam vibrations. Typically, the settling time for the beams in experimental investigations was reduced by a factor of four for beams with clamped-free boundary conditions, and the damping factor was increased by 150% for beams featuring clamped-sliding boundary conditions. Typical results for smart beams featuring embedded piezo-electric actuators are presented in Figures 4.36–4.39.

Combinations of piezo-electric sensors and actuators have been employed as active components of a vibration isolation system. Experimental investigations have also been undertaken on large space structures by employing piezo-electric actuators and sensors to control the elastodynamic behavior. The damping of the truss system was increased by a factor of 10 whereas the mass of the structure was increased by only 3 percent, as shown in Figure 4.39.

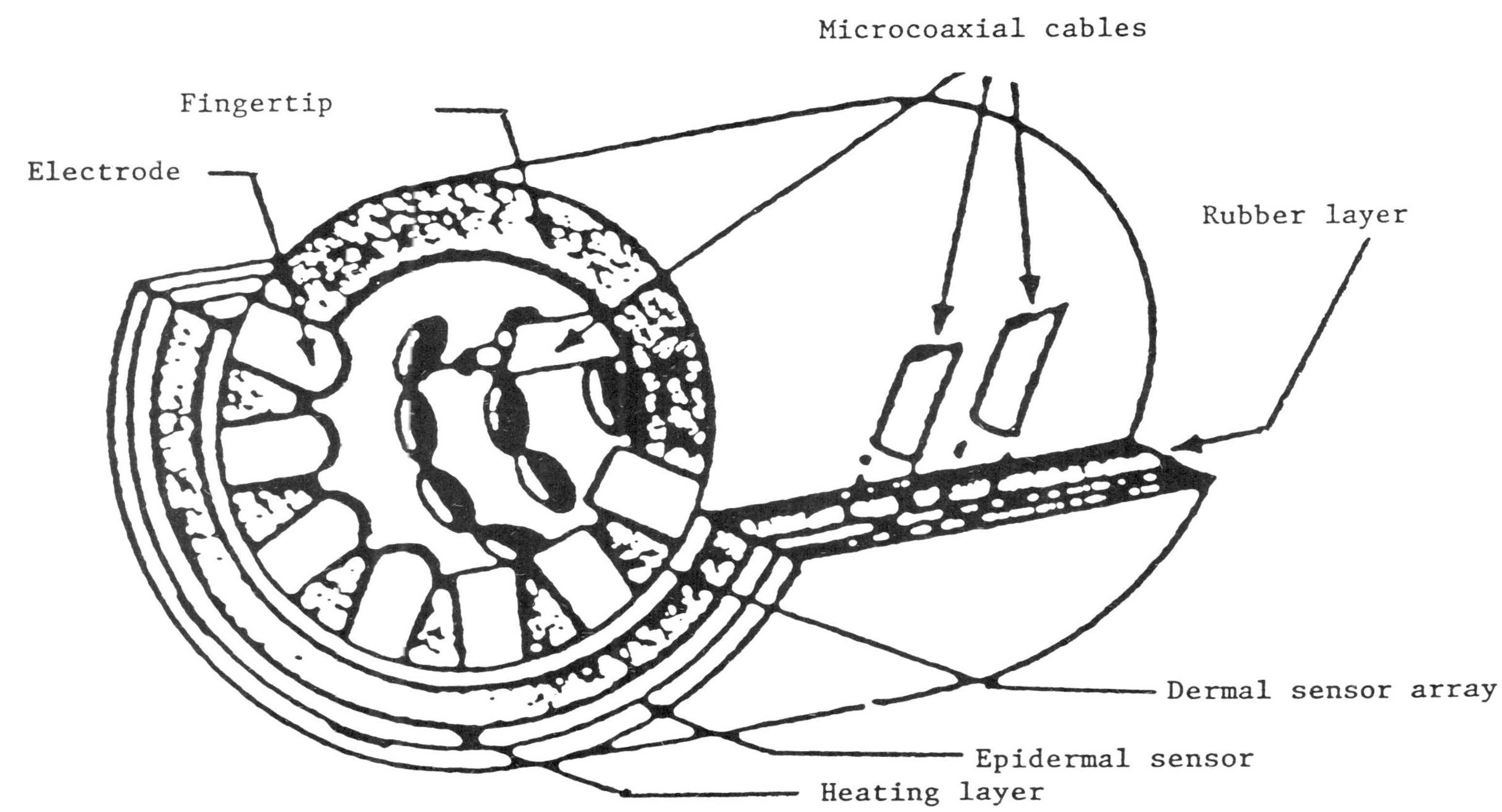

FIGURE 4.34 A portion of the University of Pisa robot finger.

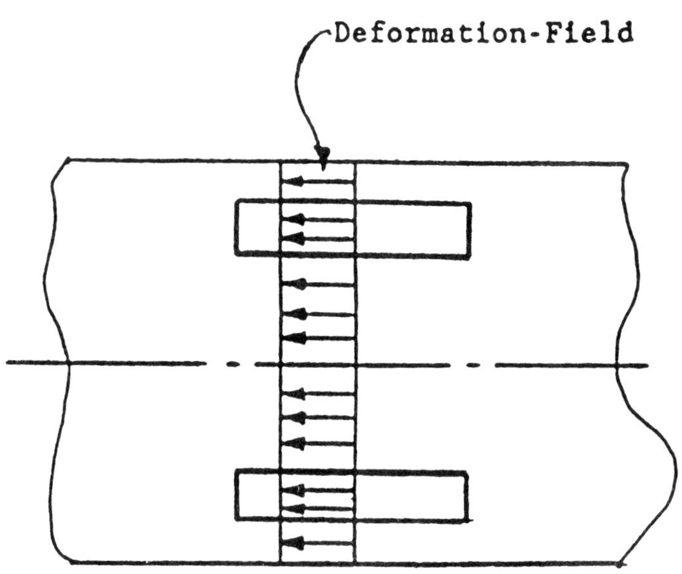

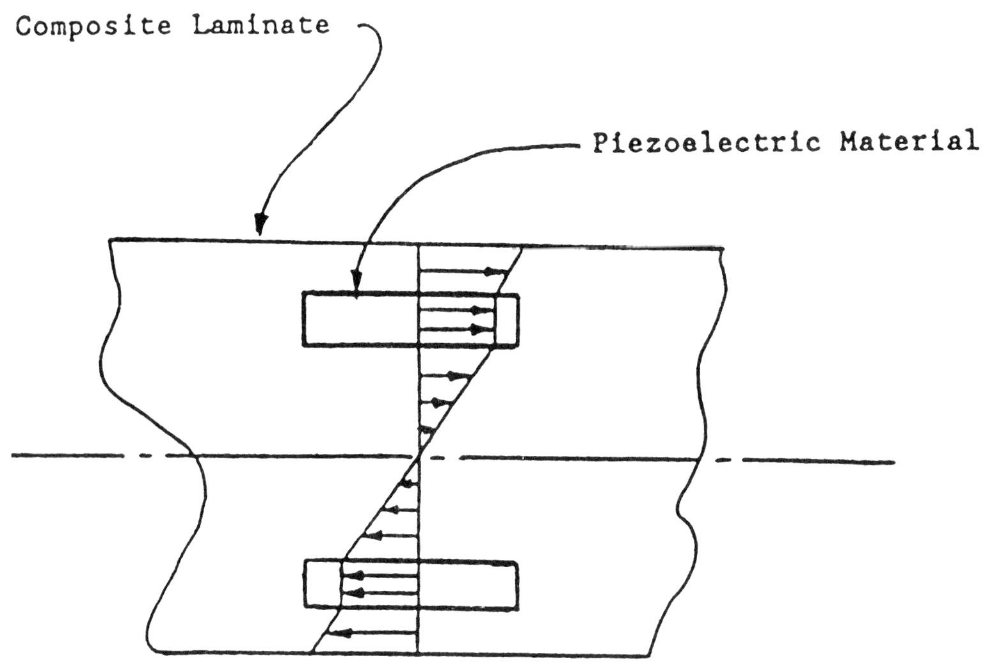

FIGURE 4.35 Smart composite laminate with embedded piezo-electric actuators.

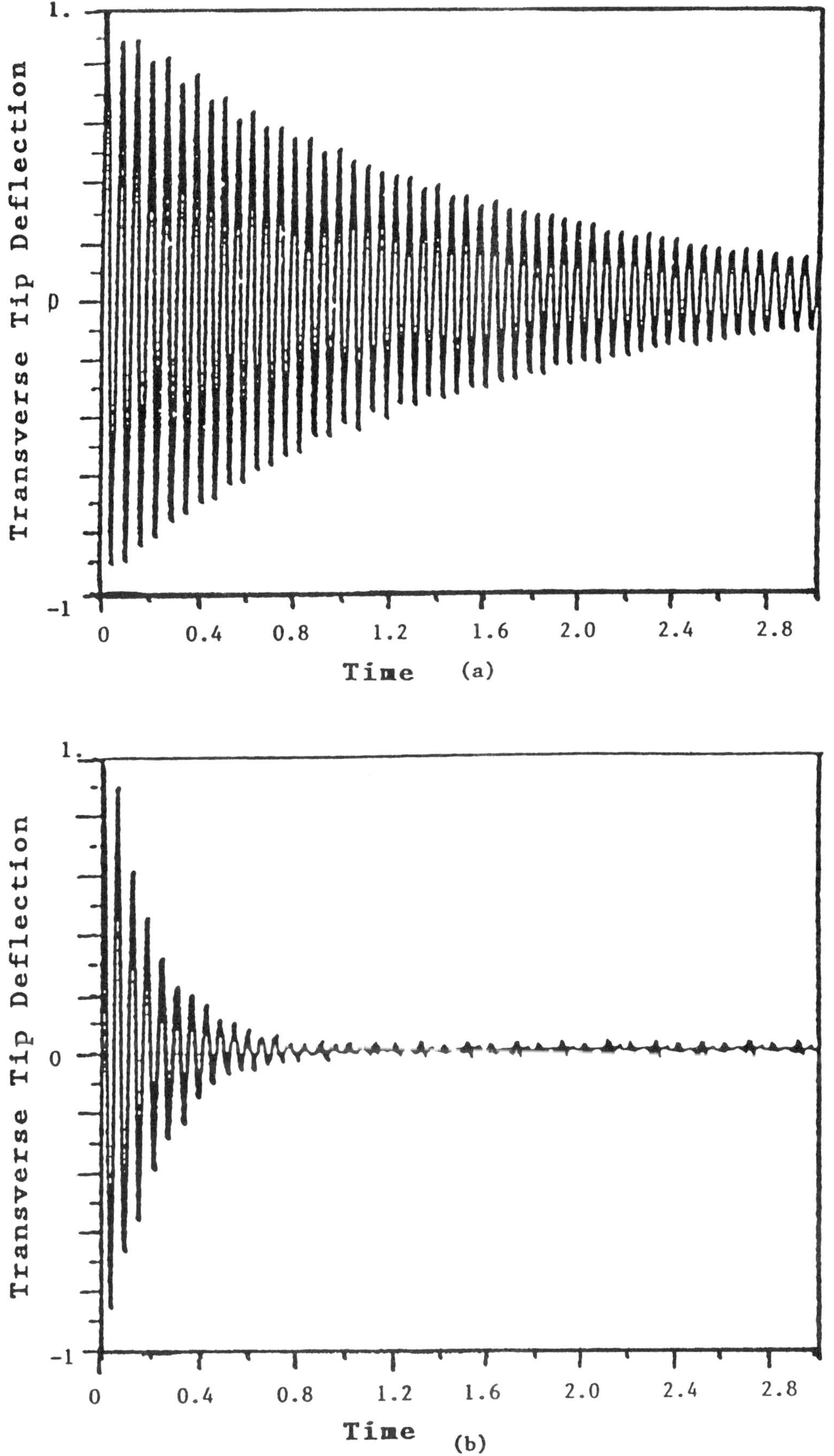

FIGURE 4.36 Control of smart beam featuring embedded piezo-electric actuators: (a) first bending mode free decay; (b) first bending mode decay with rate-feedback to all actuators.

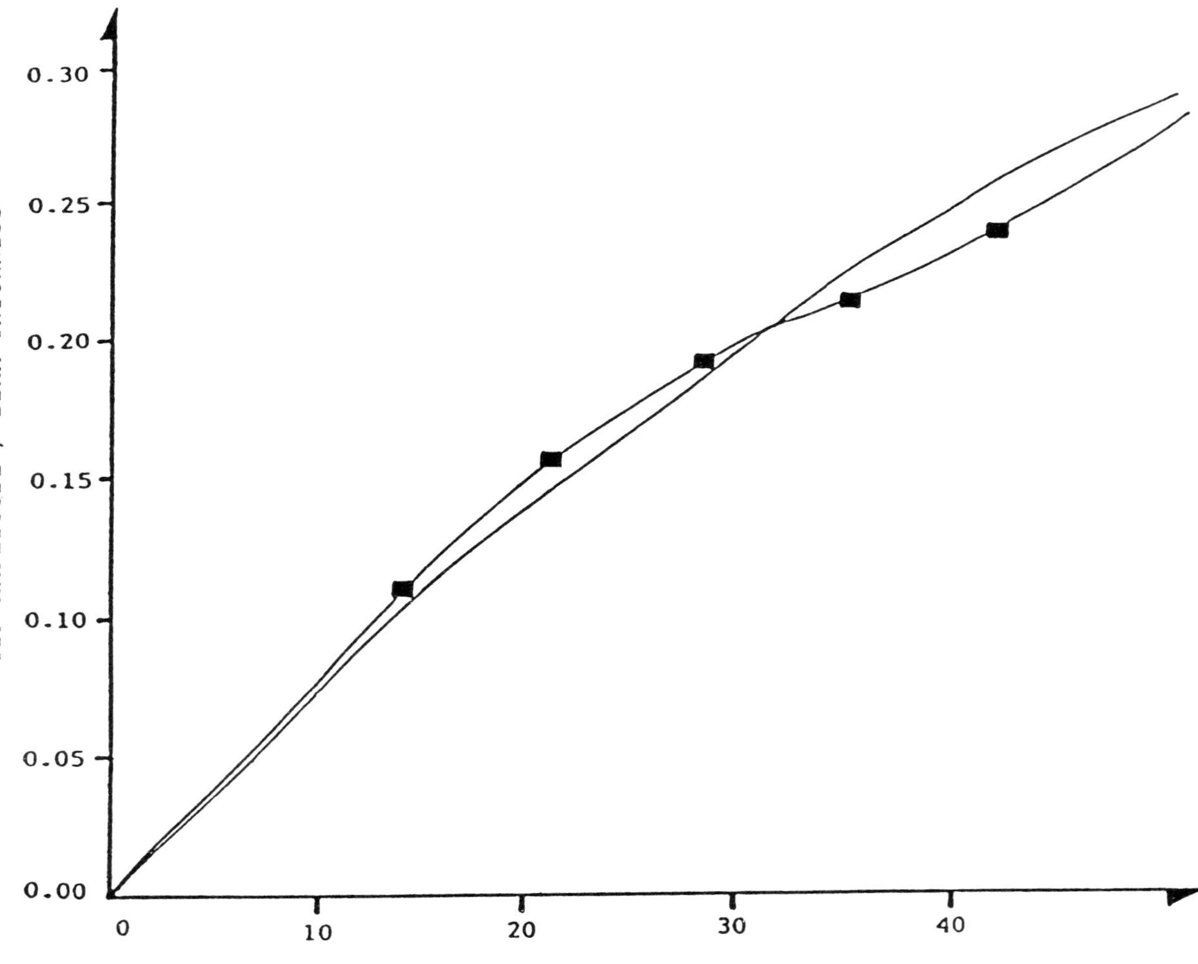

FIGURE 4.37 Experimental and analytical results for second mode bending of a smart-beam featuring embedded piezo-electric actuators.

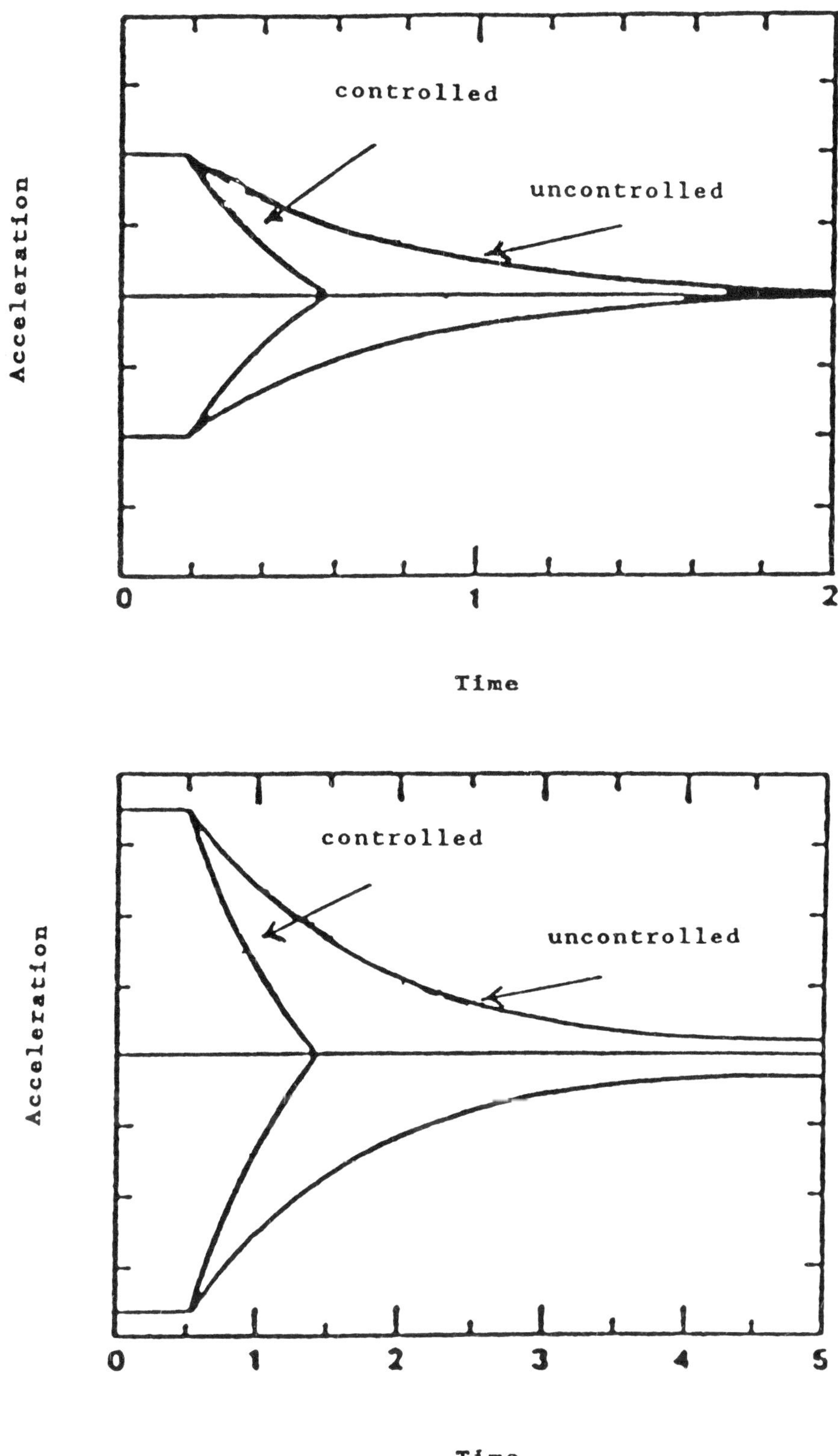

FIGURE 4.38 Acceleration response envelope of a smart beam featuring embedded piezo-electric actuators (mode 1 and mode 3 for a simply supported beam).

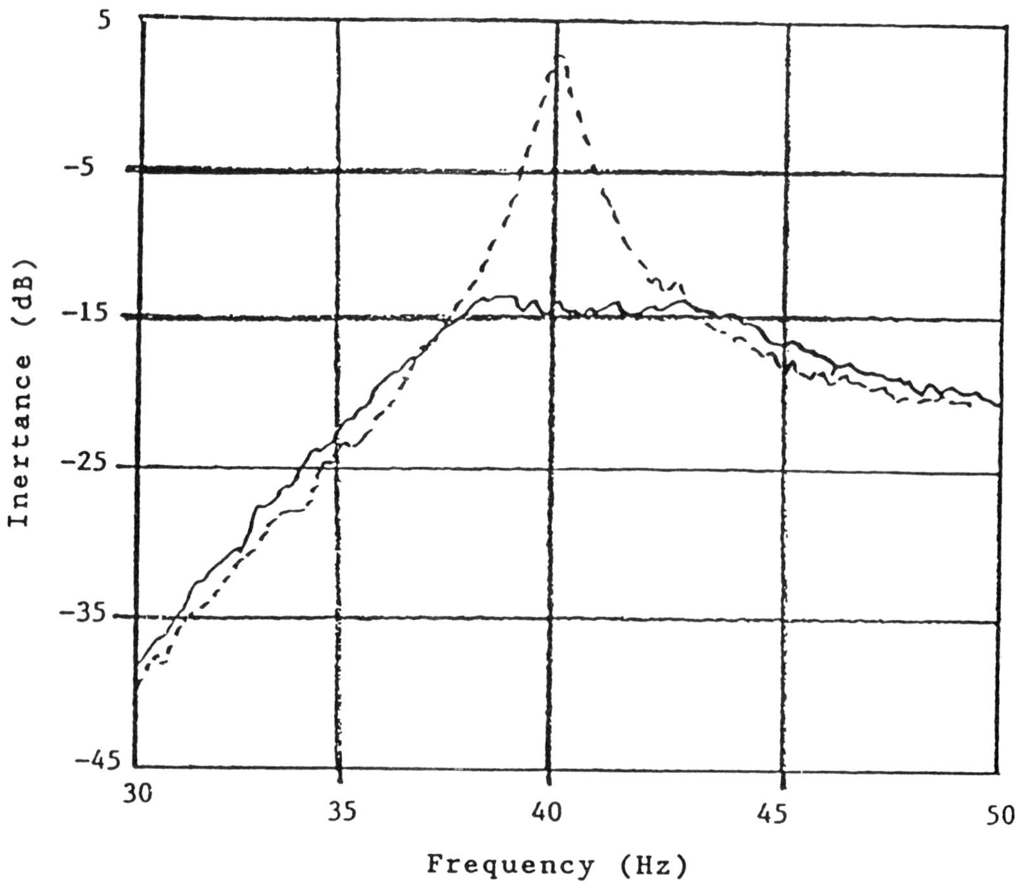

FIGURE 4.39 Experimental first mode transfer function for an aluminum truss damped by a piezo-electric proof mass damper.

4.4 Fiber Optics

Fiber optic sensing is a young technology based simply on altering the characteristics of light in response to various stimuli. This class of sensing systems offers very distinct advantages that include increased sensitivity over existing techniques; geometric versatility to be configured in arbitrary shapes; a common technology base from which devices can be constructed to sense various physical perturbations of acoustical, magnetic and thermal nature; and dielectric construction which permits these devices to be deployed in high voltage, electrically noisy, high temperature, and corrosive environments. Other advantages include:

- freedom from influence by external electromagnetic disturbances
- immunity from crosstalk
- signal flow is unidirectional if desired
- no problems with ground loops and offset dissimilar voltages where conductors meet
- very high data transmission rates up to several GHz
- simple signal multiplexing by a variety of means

- reduced costs for equivalent transmission capabilities
- lower losses and less electrical power consumption
- a high degree of security against tapping into signal trains
- greatly reduced electrical hazards and no problems with arcing or sparking
- highly resistant to adverse environmental conditions

All fiber optic sensors comprise three major systems: the optical transmitter system, the optical modulator system and the optical receiver system as schematically presented in Figure 4.40. The transmitter essentially creates an appropriate light signal that is conducted into the fiber and transmitted to the actual area of external stimulus (strain, temperature, acceleration, position, etc.).

It is here that the stimulus modulates the amplitude, phase, color or polarization characteristics of the light beam. This modulated light must now be detected and analyzed by the receiver to yield the fullest possible information regarding the stimulus.

Fiber Optic Sensors

VARIABLE	METHODOLOGY
Force	Induced birefringence
Pressure	Piezoelectric effect
Bending	Piezoabsorption
Density Change	Luminescence
Electric Field	Electro-optical effect
Dielectric Polarization	Electrochromatism
Electric Current	Electroluminescence
Magnetic Field	Magneto-optical effect, Faraday effect
Magnetic Polarization	Magnetoabsorption
Temperature	Thermal change in refractive index, absorptive properties, or fluorescence, thermoluminescence
Photoelectric Emission	Fiber defects leading to alteration in refractive index and absorptive properties
X-rays, Gamma rays	Radiation-induced luminescence
Changes in chemistry	Changes in absorption and refraction index composition due to chemical effects, chemoluminescence

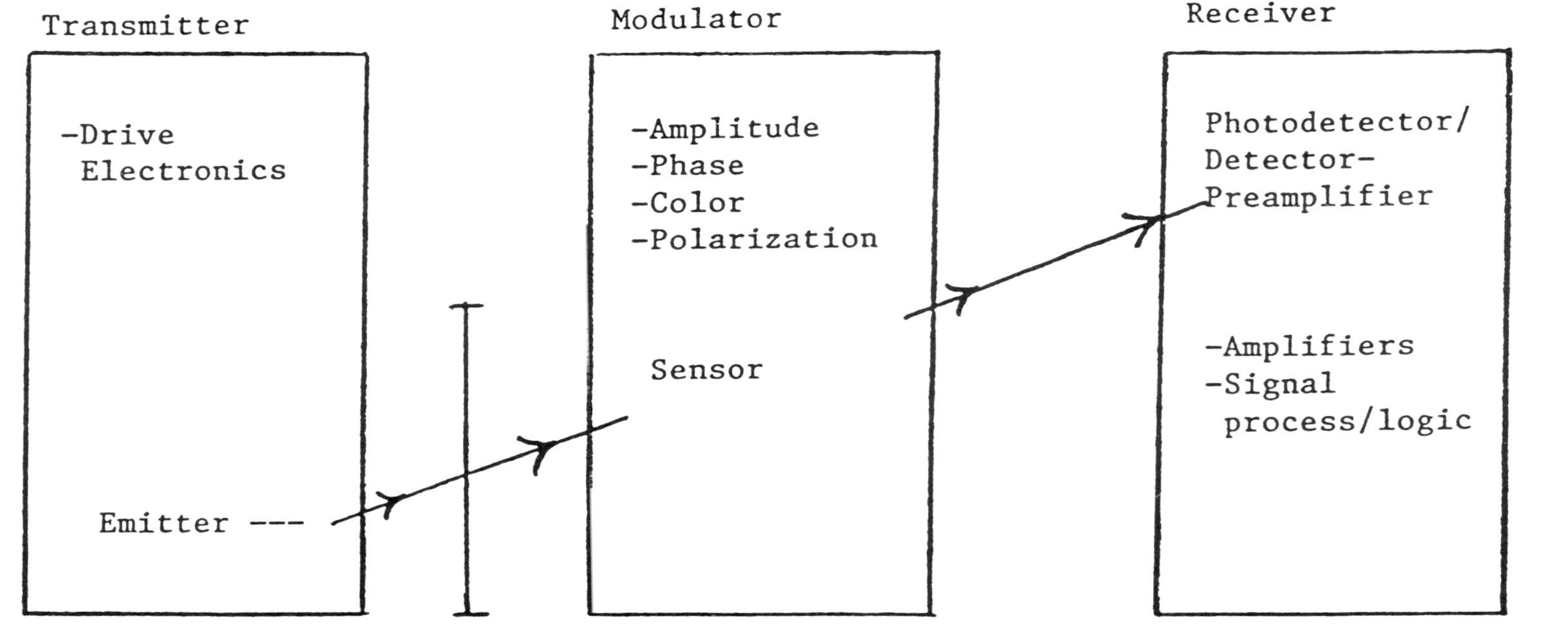

FIGURE 4.40 A typical fiber optic sensing system.

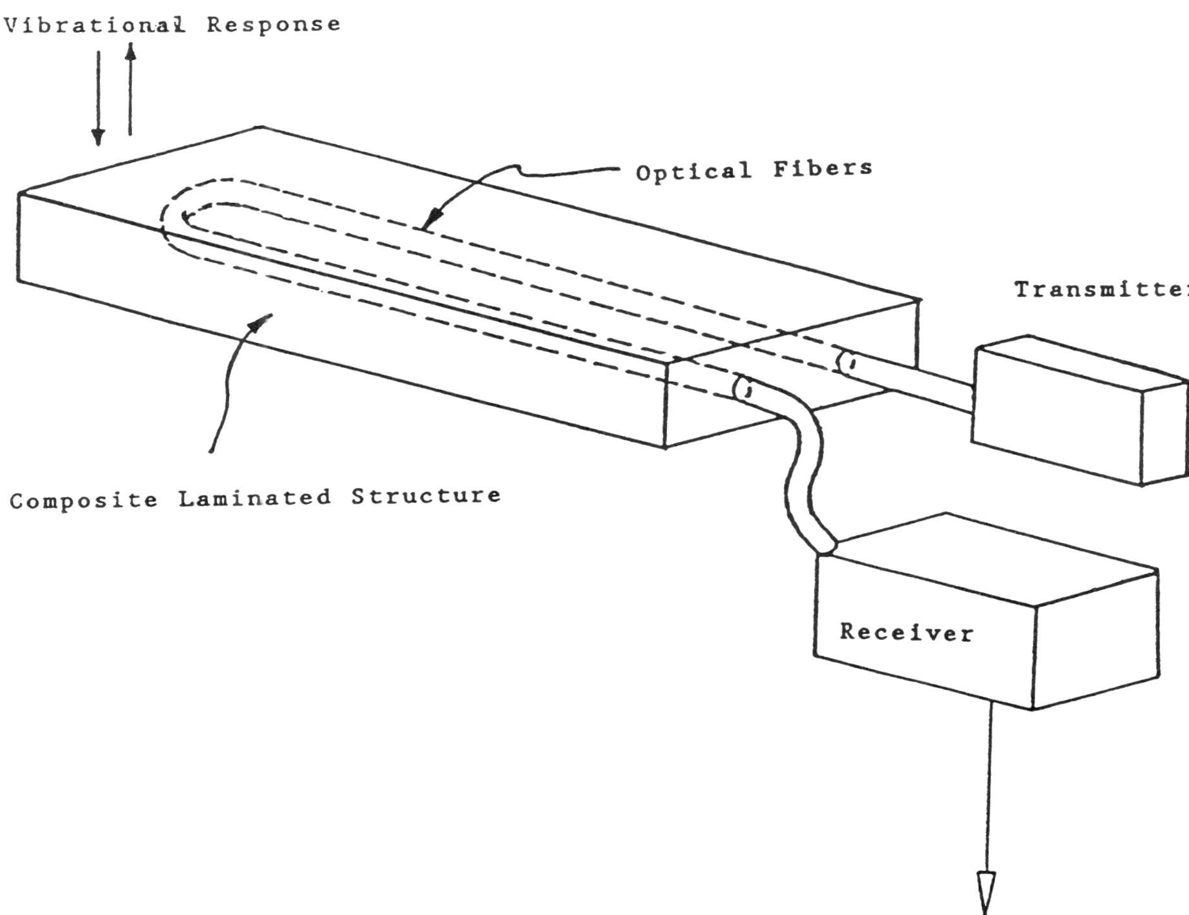

FIGURE 4.41 Embedded fiber optic sensing system for monitoring vibrations.

Fiber optic sensors are characterized as amplitude or interferometric sensors. In the amplitude sensors, the perturbation interacts with the fiber to directly modulate the intensity of light in the fiber. These sensors feature simplicity of construction and are compatible with multi-mode fiber technology with some sacrifices in sensitivity. Since most applications do not demand extremely sensitive sensors, the market for amplitude fiber optic sensors is tremendous.

Interferometric fiber optic sensors provide several orders of magnitude increased sensitivity as compared to existing technologies. This class of sensors is ideal for markets where geometric versatility and extremely high sensitivity are the dominant requirements.

An innovative class of fiber optic sensors for strain measurement is being patented by the Intelligent Materials and Structures Laboratory at Michigan State University. These sensors employ optical fibers of different diameters in order to integrate the significant advantages associated with both intensity-based and interferometric optical sensors.

Although tremendous progress has been made in optical fiber and sensor technologies, the technologies are not yet fully developed for the ultimate commercial exploitation. Practical problems need to be resolved in the areas of noise sources, detection processes, packaging and optimal fiber coatings. Ultimately fiber optic sensors must be robust enough to sense perturbances with high fidelity in the presence of external noise sources under stringent environmental conditions in typical deployment situations.

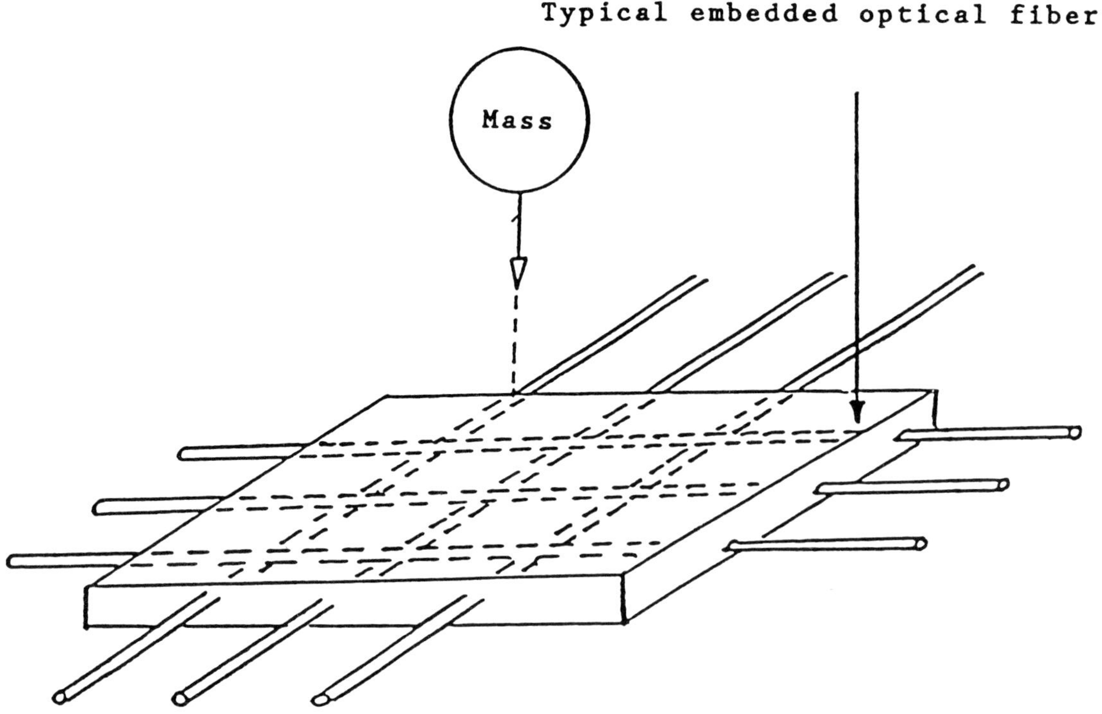

FIGURE 4.42 Embedded fiber optic sensors for detecting impact loads.

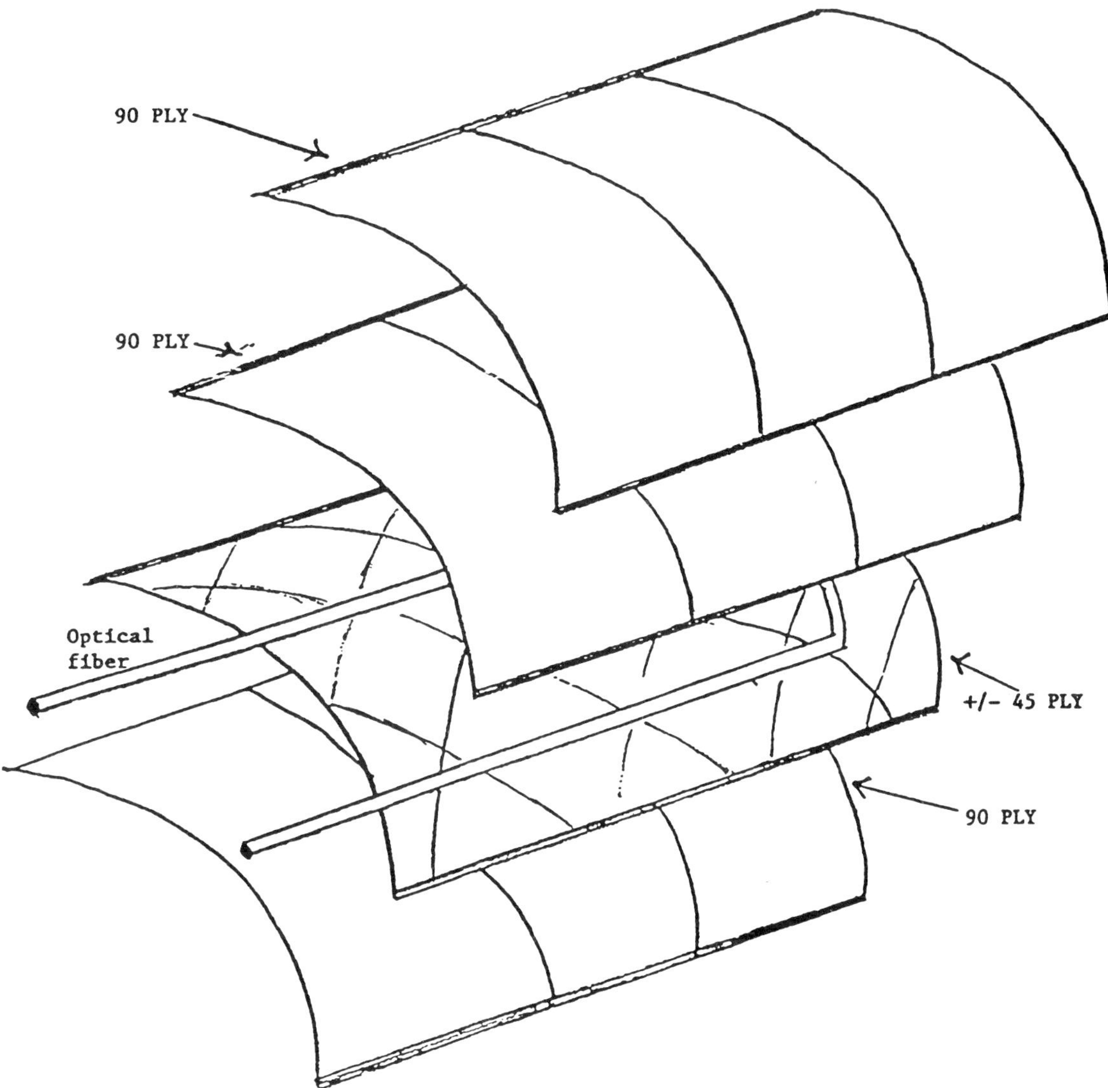

FIGURE 4.43 Embedded fiber optic sensors employed in shell-like configurations.

Continuum Applications

Sensors based on fiber optics technologies are excellent candidates for incorporation in smart structures because of their smaller size, large bandwidth, lower power consumption, geometrical flexibility, and low weight. Furthermore, whereas they may be bonded to the surface of the structure, they offer the distinct advantage of being able to be embedded within the fibrous composite materials which are frequently employed in smart structures. This feature permits the health of the structure to be monitored at all stages of its life. Thus, the critical cure properties during manufacture can be carefully monitored to ensure that the structural members attain the desired mechanical characteristics on birth. The embedded sensors also permit the health of the structure to be monitored during service, and the onset of failure can be detected thereby averting a catastrophic death.

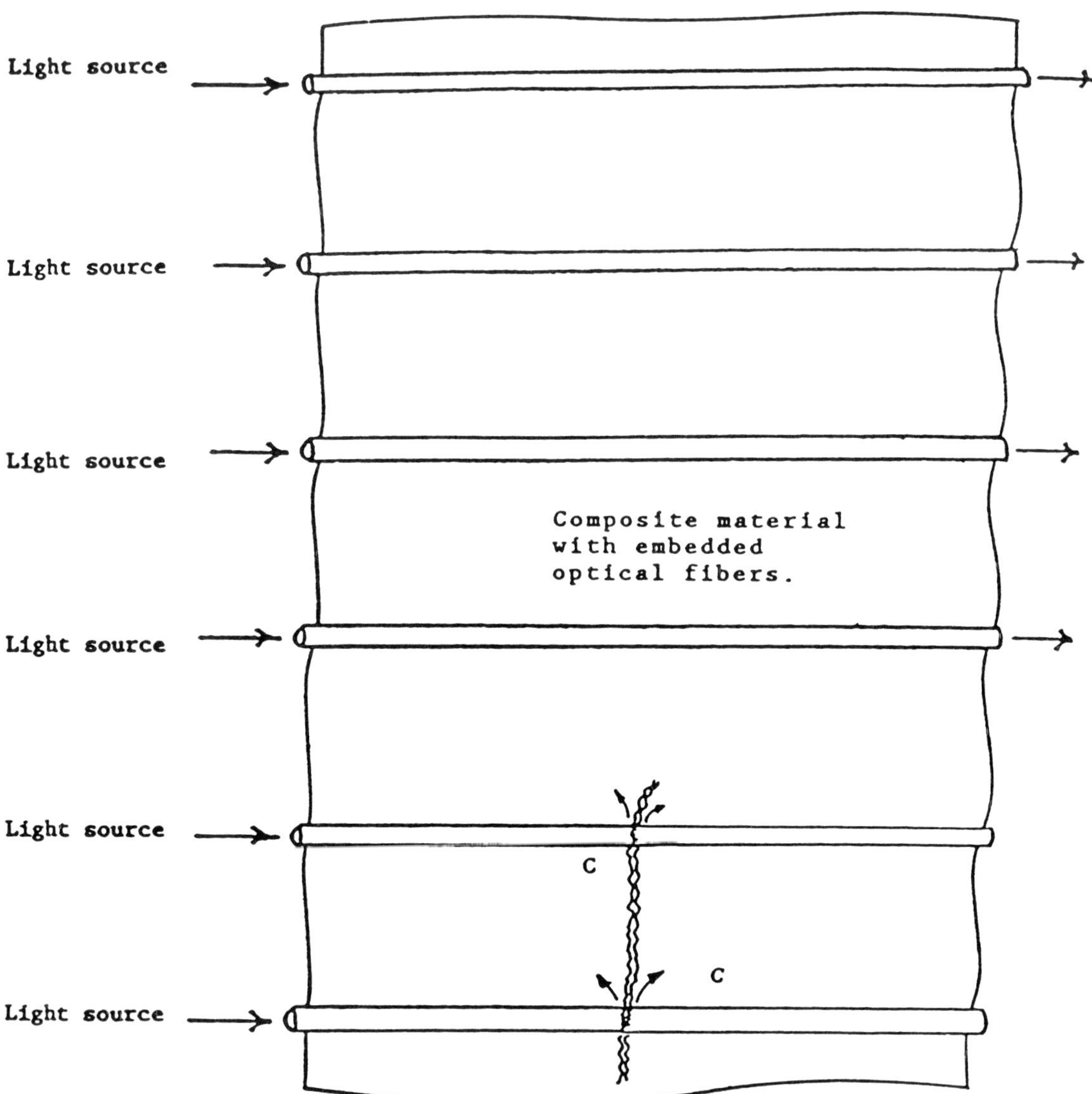

Embedded optical fibers break at the crack and prevent the light source from being received at the detector.

FIGURE 4.44 An embedded optic fiber sensing system for damage detection.

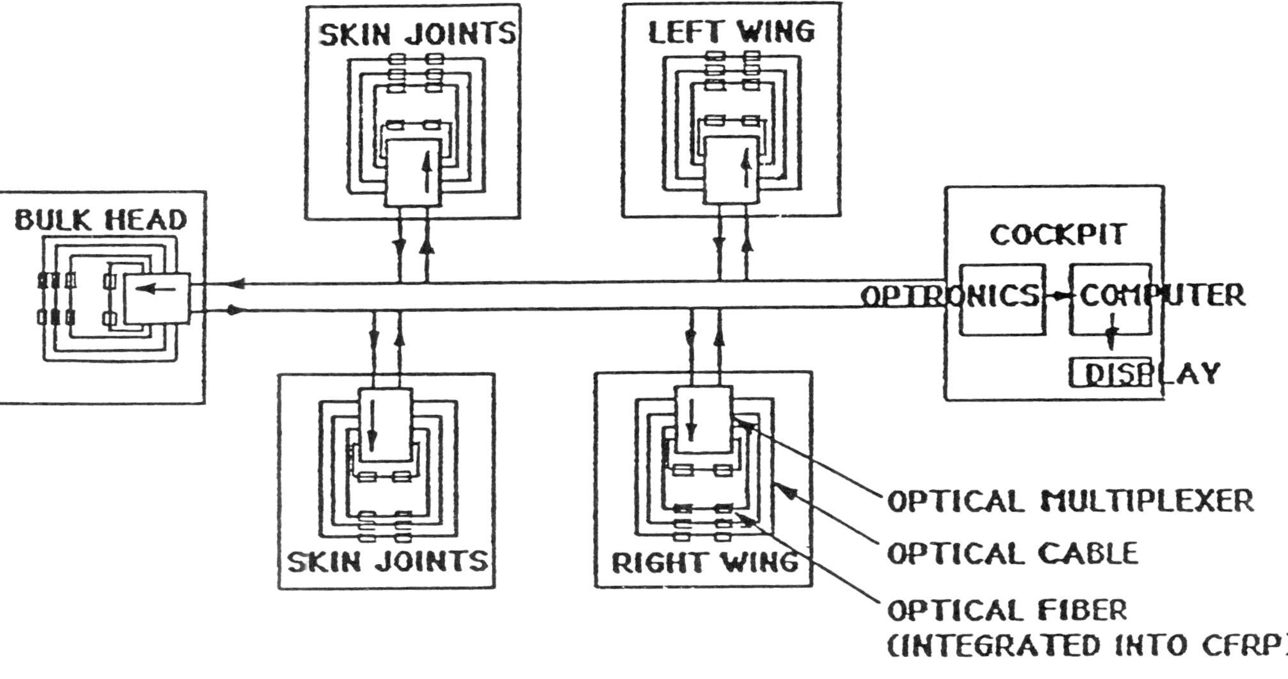

FIGURE 4.45 In-flight structure surveilliance would require fiber optic nervous system (FONS) channels in key structural regions of a plane.

Research on smart cantilevered beams incorporating fiber optic sensors has yielded promising results in the context of monitoring the dynamic behavior of the system. A scheme for monitoring the vibrational response of structures is presented in Figure 4.41. Figure 4.42 presents a schematic diagram of research involving impactive loads on smart plate-like structures.

Fiber optic strain measurements in graphite/epoxy filament-wound tubes have also been successfully undertaken. A schematic diagram of a typical fabrication is shown in Figure 4.43.

A schematic of a fiber-optic array embedded in an advanced composite material to detect damage is shown in Figure 4.44. This strategy enables critical stress levels to be identified by monitoring catastrophic fiber optic breakage such as a points C in the structure.

In-service monitoring systems that detect damage can save the aerospace industry large sums of money by eliminating periodic inspections and catastrophic structural failure. These technologies are being developed in the U.S. by McDonnell Douglas, United Technologies, and Hercules under the BRITE Initiative coordinated by Bertin in France and in West Germany by MBB under the FORS fiber optic crack-detection initiative.

McDonnel Douglas's fiber optic system employs a laser diode in conjunction with a loop of optical fiber and a photodetector. The diode serves as a source of laser light that is transmitted through the optical fiber to the photodetector. This system employs a Corning Glass Works microbend-sensitive type polyimide-coated fiber that has an external diameter of 125 μm and an inner core of diameter 8 to 10 Mμ. Systems of this kind can easily measure strains as low as 10^{-12} μ inch/inch and can be employed to determine the strains at critical points in a structure, which may be as high as 50 in an F-15 aircraft, for example.

Future developments in this area may focus on the development of fiber optic nervous systems (FONS) in the structure of aerospace systems as shown in Figure 4.45, and these FONS would have capabilities to undertake such in-flight structure surveillance as assessment of battlefield damage, measurement of temperature to recognize ice on the wings, and performance of the powerplants. The ability to undertake this multitude of tasks in real-time will have a tremendous impact on the design, engineering, servicing and economics of the aerospace industry worldwide.

5 TECHNOLOGICAL DEFICIENCIES

The smart materials and structures technologies discussed in this intelligent report are in transition from embryonic to emerging technologies through vigorous research efforts in the industrial, commercial and academic environments. The marketplace topology for this impending revolution is highlighted in Chapter 6, and projections for market growth are presented in Chapter 7.

Whereas the tremendous potential of these innovative and revolutionary technologies may not be fully realized for another twenty years, pioneering products based on these technologies should start appearing in the marketplace as early as 1992/1993. Before a company makes a strategic decision to enter this high-stakes business in order to develop a strong market position, it is important that the principal technological impediments to the evolution of these technologies be fully recognized. Some of these technical impediments are presented in the subsequent paragraphs.

Electro-Rheological Fluids

Stability

The stability of electro-rheological (ER) fluid suspensions has not been resolved for several classes of fluids. Thus, if an ER fluid device is not activated for a prolonged time-period, the particulate phase settles to produce a non-homogeneous suspension. This problem is a frequent one with smart skins where thicknesses of the fluid domains are generally small. Various surfactants have been employed to minimize this stability problem.

Thermal Effects

The power necessary to activate certain classes of ER fluids increases with temperature, which has an impact on cost-effective, compact power sources. In addition, if the ambient operating temperature is high, then hydrous fluids are not viable candidates for certain classes of applications. Water-based fluids can be successfully deployed typically in the temperature range $-15°C$ to $75°C$. This limitation has fueled the search for viable anhydrous fluids.

Electrode Separation

ER fluids possess characteristics that are dependent upon the field strength of the electrical field imposed on the fluid domain. Typically this field is measured in kV/mm. Thus, in order to tailor the rheological behavior of the fluid, the electrode separation is typically of the order of a few millimeters in order to limit the power requirements. Electrodes must be carefully designed for high voltage applications, in order to avoid arcing with the associated degradation of the electrode material.

Certification

The application in which an ER fluid will be deployed dictates the criteria that will be imposed on its composition. For example, fluids employed in commercial aircraft must comply with FAA regulations, fluids employed in aerospace systems must be space-qualified, and fluids for automotive products must comply with criteria imposed by the NHTSA and EPA.

Interaction

The chemical reaction between the ER fluid and the containment material must be carefully evaluated for certain classes of fluids in order to assure the customer of a long trouble-free service-life.

ER Phenomenon

The fundamental ER-pseudo-phase change phenomenon obtained upon applying an electric field to the ER fluid is not understood. Several theoretical models and hypotheses have been propounded, but there is no universal agreement on the basic mechanism to describe the ER-effect.

Continuum Applications

Traditional ER fluid applications have been discrete in nature, such as fluid valves and clutches. In these applications design rules are scarce. The knowledge base is even worse in the embryonic field of continuum applications where ER fluids are embedded in structural materials. Design rules need to be established in order to synthesize structures with specified performance characteristics. Typical design parameters are the size and shape of the fluid domains, the roughness of the structural material at the fluid-structural interface, the fluid characteristics for a specified structural material, electrode discreteness, and the thermal environment.

Control Issues

The area of control strategies to be employed in the deployment of ER-based smart materials is an open field, involving sensor selection, sensor location, datalinks, microprocessors and control methodologies.

Piezo-Electric Materials

Strength Reduction

The embedment of piezo-electric materials within a smart structure introduces undesirable stress concentrations that lead to weaknesses. In some applications the piezo-electric material can be bonded to the surface of the structural material to avoid reducing the strength and fatigue life of the parts.

Thermal Effects

The vulnerability of piezo-electric actuators to extreme service temperatures and thermal cycling is not clearly understood at this time; therefore, the domain of applicability of this class of actuator and sensing systems is not clearly defined.

Certification

Smart materials based on piezo-electric actuation and sensing systems must satisfy the stringent FAA regulations in aircraft applications and space qualification criteria for space applications. Automotive applications of this technology must comply with NHTSA requirements.

Interaction

The chemical reaction between the piezo-electric sensing and actuation materials and the host structural material must be carefully addressed in order to guarantee viable performance under variable service conditions.

Continuum Applications

Traditional applications of piezo-electric actuators and sensors have been restricted to discrete applications. The knowledge base for continuum applications, where layers of piezo-electric materials are embedded in structural materials, is embryonic in nature. Additional research and development is required in order to establish design rules in order to synthesize viable cost-effective structures and prescribed performance characteristics.

Controls Issues

Control strategies for this class of smart materials and structures are still in the embryonic stage of development. Several theoretical, software and hardware issues must be addressed prior to commercialization of piezo-electric-based smart materials technologies.

Shape Memory Alloys

Slow Response-Time

The response-time of smart materials featuring shape memory alloys is greater than for their competing technologies due to the inherent thermal time constants.

Hysteresis

The hysteresis characteristics of shape memory alloys must be carefully evaluated and clearly understood in order to ensure desired performance and repeatability under variable service conditions.

Energy

The energy required to activate the shape memory phenomenon is significantly larger than other approaches for certain classes of problems because of the attendant thermal losses, primarily by conduction.

Temperature Constraint

The shape memory phenomenon occurs at a single specific temperature for a prescribed class of materials. Hence, in order to activate the material, the shape memory alloy must be heated to the prescribed transition temperature at the boundary of the martensitic and austenitic domains of the phase diagram. This single actuation temperature could be problematical, if the material is required to operate in unstructured thermal environments.

Stress Concentration

The embedding of metallic wires in a host material will result in stress concentrations at the interface between the two constituents when the shape memory alloy wires are activated. These stress raisers will clearly have an undesirable effect upon the structural integrity of the part and its fatigue life.

Certification

Smart materials based on shape memory alloys must satisfy the stringent FAA regulations in aircraft applications, and space qualification criteria for space applications. Automotive applications of this technology must comply with NHTSA requirements.

Interaction

The chemical reaction between the shape memory alloy and the host structural material must be carefully addressed in order to guarantee viable performance under variable service conditions.

Continuum Applications

Traditional applications of shape memory alloys have been restricted to discrete applications. The knowledge base for continuum applications, where layers of shape memory alloys are embedded in structural materials, is embryonic in nature. Additional research and development is required in order to establish design rules in order to synthesize viable cost effective structures with prescribed performance characteristics.

Control Issues

Control strategies for this class of smart materials and structures are still in the embryonic stage of development. Several theoretical software and hardware issues must be addressed prior to commercialization of shape memory alloy-based smart materials technologies.

Fiber Optics

Integration Issues

The integration of optical fibers and a host material create several integration issues. For example, most fiber optic cables are unable to withstand the cure temperatures of approximately 350°F associated with many of the aerospace epoxy resin systems. Fortunately, sheathed fibers are available that do not suffer from this limitation.

Sensitivity

The sensitivity of the fiber optic sensing system is dependent upon the phenomenon being measured. Whereas considerable progress has been made in the transmission of data by optical fibers, the development of fiber optic sensing systems is certainly an embryonic technology.

MARKETS AND APPLICATIONS

Smart materials and structures technologies presented in this intelligence report will have a tremendous impact in reshaping the technological and economic bases of the international business environment during the next two decades. A detailed description of the segments of the marketplace which will be revolutionized the most is presented in Chapter 6. In addition to the specific technical applications discussed in Chapter 6, it is crucial to recognize that significant order-of-magnitude payoffs will arise in various applications as a consequence of the implementation of smart technologies. Some of these payoffs are highlighted below.

Technological Impact

- Technological developments that are considered impossible today will become routine realities as a result of smart technologies, e.g., National Aerospace Plane.
- Completely unforeseen markets will be created and the dominant players will carve out distinct segments of these markets.
- The design of smart parts and subassemblies for smart mechanical and structural systems will be simplified, and significant order-of-magnitude advantages will accrue due to design-cascading effects.
- Smart technologies will result in simplified and efficient design and manufacturing processes.
- Performance characteristics of smart mechanical and structural systems will not only be significantly superior to today's systems, but also optimal for variable service conditions and unstructured environments.
- Significant reduction in costs will accrue due to the integration of sensing, actuation and damage or quality monitoring functions in smart structures.
- Maximum utilization of microprocessor technologies due to the unique ability of smart materials to interface with modern solid state electronics will permit the quantum leaps in the electronics area to be immediately transformed into technological advances in the smart materials and structures arena.

- Smart structural and mechanical systems will exhibit significantly superior response characteristics and capabilities due to reduced inertia, as compared with conventional technologies.
- Smart materials and structures will provide designers with a unique capability in the history of mankind to optimize system performance under various service conditions and unstructured environments.

Economic Impact

It is clearly evident from the market topology presented in Chapter 6 that the impending revolution in smart technologies will impact every segment of the world marketplace. Therefore, the economic impact of these technologies on the international marketplace is very significant. There is no such thing as the "best smart technology," and no one is in a position to definitively assess which technology is the best for various specific applications. One can project that as the smart technologies field matures, individual technologies will be integrated to yield hybrid smart technologies that will permit the stringent performance specifications of various smart part and subsystems to be adequately satisfied. Therefore, it is somewhat premature to project the market share and the market growth associated with each of the key smart technologies presented in this intelligence report. However, at this stage in the evolution of smart technologies, it is more appropriate to provide projections for the integrated market share and market growth for all these technologies combined. This philosophy has been adopted to project the market growth for smart materials and structures in Chapter 7.

Smart materials and structures technologies will significantly impact every conceivable sector of the marketplace. The economic and technological impact of these technologies is discussed in Chapter 6 and the projected market growth is presented in Chapter 7. Hy-Tech International believes that the most significant impact of these technologies will be in the following sectors of the economy:

- automotive and transportation industries
- aerospace industry
- defense industry
- biomedical devices
- advanced manufacturing, robotics, and industrial machinery
- consumer products and sporting goods
- high precision instruments, printed circuit boards, and electronic packaging
- highways, buildings, and bridges

Specific applications in each of these market sectors will be motivated by substantial order-of-magnitude improvements in performance characteristics due to the implementation of smart materials and structures technologies for variable service conditions and unstructured environments. Economics will be the other driving force in the implementation of smart technologies since the cost of these smart products will be substantially lower due to a smaller number of moving parts associated with smart technologies, the ability of smart materials to interface with opto-electronics, and several other design cascading advantages at the system level.

Typical applications of smart materials and structures technologies in each of these sectors of the economy are highlighted in the following pages.

Automotive and Transportation Industries

- engine mounts (minimize vibration and noise)
- active suspensions (improve ride and handling)
- steering systems (improve system performance)
- shock absorbers
- smart windshields
- smart bumpers (improve crashworthiness)
- pumps, valves and actuators (improve system performance)
- clutches and transmission systems (smooth start-up and improved performance)
- anti-lock braking systems
- new generation of engines
- automotive springs

Aerospace Industry

- smart skins containing phased-arrays (to permit aircraft to sense and communicate in various frequency bands and in any direction)
- smart wings (improve manufacturing quality and productivity, improve aerodynamics and system performance in service, detect damage and impending failure)
- smart control surfaces
- smart rotor-craft systems (minimize vibrations, improve system performance under variable service conditions)
- vibration suppression systems
- instrument panels (improved ergonomical design for strategic and tactical systems)
- undercarriage, shock absorbers, landing gear
- hydraulic pumps, actuators and valves
- missiles
- large space structures (improve performance in unstructured environments)
- commercial aircraft (reduce drag for fuel efficiency, minimize noise and vibration)
- space robots
- self-deploying space structures

Defense Industry

- submarines (reduce drag, quieter submarines, quieter weapon-launch systems)

- surface vessels (reduce drag, quieter)
- hydraulic valves, actuators and pumps
- switches
- variable radar and acoustical signature
- smart skins (life through death health monitoring capabilities: manufacturing, service, battlefield damage)
- ammunition supply systems
- materials handling equipment
- smart armour
- SDI (space structures supporting weapons and antenna for retargetting maneuvers without detrimental jitter and thermal-flutter)
- stealth technologies (to evade enemy emitter and platform identifications to make more specific threat determinations)
- Field-repairable airframe structures (innovative high-lift rotors for helicopter super-maneuverability)
- acoustically-damped torpedo propellors

Biomedical Devices

- prostheses: artificial limbs, hands and joints
- orthodontic braces
- sclerosis
- wheelchairs
- treatment of sporting injuries (fractures)
- implants

Advanced Manufacturing, Robotics, and Industrial Machinery

- active balancing
- dynamically-tunable robot arms
- noise reduction
- vibration control of machine-tool structures (improve quality and productivity)
- hydraulic actuators, valves and pumps
- joint actuators for articulating robotic systems
- robotic end-effectors
- smart flexible fixtures and grippers (to handle workpieces of various shapes and sizes)
- material handling
- farm equipment
- oil drilling and mining equipment (improved performance in unstructured environments)
- textile machinery

Consumer Products and Sporting Goods

- skis
- tennis rackets

- golf clubs
- fishing poles
- baseball bats
- snowmobiles
- bicycle industry
- switches
- washing machines
- household dryers
- vacuum cleaners
- lawn mowers
- snow blowers

High Precision Instruments, Electronic PCBs, and Packaging

- advanced vibration isolation concepts
- dynamically-tunable optical characteristics
- dynamically-tunable thermal characteristics

Highways, Buildings, Bridges

- smart foundations
- smart skins
- dynamically-tunable optical and thermal characteristics for doors and windows (energy efficiency and comfort)
- smart structures, bridges, buildings (minimize vulnerability to damage)
- building elevators

7

MARKET PROJECTION AND GROWTH

Smart materials and structures technologies will revolutionize a broad segment of the international marketplace for products in the defense, aerospace, automotive and commercial-products industries. The projections for the market growth in those sectors that will be impacted the most are highlighted in the following sections. The total marketshare of smart materials and structures is projected to exceed $65,000 million by the year 2010.

The aerospace industry and the defense industries will be the first to significantly fully capitalize on these smart technologies. The ultimate applications of smart technologies in these industries may not be realized until after the year 2000 when the high end of these high-tech smart technologies will have been fully exploited. However fiber optic-based smart skins will be seen on a limited basis on prototype F-15 aircrafts, for example, by 1992.

This scenario is in sharp contrast to the automotive industry, where a high-volume of low-end smart technologies will be employed in production units as early as 1995. Typically, these applications would be based on hydraulic applications utilizing ER fluids in fluid clutches for example; and the use of memory metals and piezo-electric technologies in solenoid actuators and triggering mechanisms, for example.

It is clearly evident from the Markets and Applications presented in Chapter 6 that the impending revolution in smart technologies will impact virtually every segment of the world marketplace; therefore, the economic impact of these technologies on the international marketplace will be very significant. The dust has not yet settled as to which technology is "best" for various applications; therefore, it is somewhat premature to project the market share and the market growth associated with each of the key smart technologies presented in this intelligence report. However, at this stage in the evolution of these technologies, it is more appropriate to provide projections for the integrated market share and market growth for all these technologies combined. This philosophy was adopted to project the market growth for smart materials and structures in the various segments of the economy highlighted in the following pages.

Figures in 1989 dollars have been employed to represent the total market in a given industry segment in the year 1989. However, all dollar figures projected beyond 1989 represent the dollar amounts at that point in time. For example, in the projection for the aerospace industries listed in the following tables, the total market is approximately $40,000 million in 1989 dollars whereas the projection for the commercial aircraft market in the year 2000 is $7,000 million in year 2000 dollars.

Aerospace Industries

1989		Total Market: $40,000 million
1998	hydraulic systems:	$650 million market growth 9%
2000	commercial aircraft (new aircraft & parts)	$7,000 million market growth 5%
2000	aircraft parts	$8,000 million market growth 7%
2000	military aircraft	$1,000 million market growth 12%
2010	smart devices on aircraft, space stations, spacecraft hydraulic systems	$5,000 million market growth 10%

Automotive and Transportation Industries

1989		Total Market: $180,000 million
1995	transmissions: ER-based systems	$800 million market growth 20%
1997	active suspension systems in cars and trucks	$1,000 million market growth 15%
1997	active steering systems market growth 15%	$500 million
1997	anti-lock braking systems	$500 million market growth 20%

Advanced Manufacturing, Robotics and Industrial Machinery

1997	hydraulic devices: valves, activators, pumps	$2,500 million market growth 15%
1997	robot-arms, actuators, end effectors	$500 million market growth 15%
1998	food processing	$400 million market growth 2%

1998	textile machinery	$200 million market growth 1%
2000	construction equipment	$5,000 million market growth 1%
2000	machine tools	$4,000 million
2000	mining machines	$400 million market growth 3%
2000	oil drilling equipment	$1,000 million market growth 2%
2005	farm equipment	$1,000 million market growth 2%

Biomedical Devices

1995	wheelchairs	
2000	prosthesis: artificial limbs, hands and joints	
2000	biomedical applications (combined)	$1,500 million market growth 10%

Defense Industries

1997	smart devices missiles and spacecraft	$1,000 million market growth 7%
2000	submarines, helicopters, etc.	$10,000 million market growth 5%

Highways, Buildings, Bridges

2000	highways, buildings, and bridges	$10,000 million market growth 3%
2000	U.S. government works	$5,000 million market growth 2%

Consumer Products and Sporting Goods

2000	consumer products and sporting goods	$5,000 million market growth 15%
2000	home appliances	$1,000 million market growth 12%

High Precision Industrial Packaging

1997	high precision industrial packaging	$5,000 million market growth 15%

Furthermore, $7,000 million represents the entire commercial aircraft market, only a part of which may be substituted with smart materials and structures. Similarly the market growth for the commercial aircraft market in the year 2000 is projected to be five percent per year based on the $7,000 million market in the year 2000.

Fiber Optic Sensors: Discrete Applications

Over 75 different types of fiber optic sensors have been developed in order to detect magnetic fields, acoustical fields, pressures, temperatures, accelerations, displacements, torques, currents and strains, for example. These discrete applications are responsible for driving the current evolution of the technology.

The telecommunication market in the U.S. is $700 million annually and is projected to grow at an annual rate of 25 percent. The U.S. non-tactical, military fiber optic market is $35 million. The largest tactical applications of fiber optics are for platform-command control, communication intelligence, navigation and electronic warfare systems. The fiber optic market for such sea-based platforms is currently at $10 million and is projected to increase to $100 million by 1995. Ship board installation of fiber optics is scheduled to begin in 1991, and major installation on submarines is projected to start in 1993. The deployment of fiber optics in military aircraft will be extensive and is expected to increase at a compound annual growth rate of 40 percent to reach $40 million by 1994. Ground laid cable for tactical communications and remote control, the second largest application, is currently at $5 million per year and is expected to grow at a rate of 30 percent per year. The application of fiber optics in the tether missile market is anticipated to be $40 million by the year 1995. Major missile programs include the U.S. Army fiber optic guided missile, the Navy's Sea-ray missile, the Marine Corps' advanced anti-tank weapon system and the Air-Force's GBU-15. Recently, research on a fiber optic tether for the Mak-48 Advanced Capability torpedo has been initiated. Anti-submarine warfare applications for fiber optic towed array is projected to increase at 42% and reach $15 million by 1995.

The military fiber optic sensor market is in the exploratory research stages. Fiber optic gyroscopes in aircraft, ship, satellite and missile guidance applications are expected to begin in 1991 and grow rapidly due to the low cost and long life of the gyroscope. The demand for hydrophones, accelerometers, pressure and temperature sensors is also anticipated to be significant.

The total military market for fiber optics is projected to grow at approximately 25 percent during the next decade, with growth in tactical applications to exceed 35 percent. This growth coupled with a conservatively estimated growth of 15 percent per year in the commercial sectors will make the discrete applications market for fiber optics technology the dominant market. The continuum applications market for this technology will be stimulated by this growth, but will probably continue to be a relatively minor market till the year 2000.

8

COMMERCIAL ACTIVITIES

The activities of the key companies in the area of Smart materials and structures:

American Cyanamid has decided to quit the smart materials field and is focussing its efforts in completely unrelated areas. The company is in the process of negotiating the sale and licensing of its electro-rheological fluids-based technologies.

Boeing has identified the areas of smart materials and structures and advanced avionics as its top two technological priorities. Boeing is currently evaluating all major technological options in the area including conformal electronics, which can be dovetailed with existing or next generation aircraft structures and skins.

Chrysler is very active in the field of electro-rheological fluids. Applications include suspensions, power steering and engine mounts.

De Havilland is closely interacting with the University of Toronto Aerospace Institute in order to build smart leading edges for the wings of a commuter plant. Initial prototypes are anticipated to be completed by 1990.

DuPont is evaluating the manufacturability of embedded optical fiber sensors in advanced composite materials.

Fiat has been active in the area of memory metals for discrete applications for the last six years. The company is currently exploring the possibility of developing electro-rheological-fluid-based devices for high-performance automotive and military applications.

FMC is exploring possibilities of employing smart materials and structures for submarine applications—particularly, reduction of drag and quieter launch of missiles.

Ford is carefully evaluating the potential of electro-rheological fluids for suspensions, engine mounts, power steering, and other discrete applications.

General Motors has already developed smart engine mounts based on electro-rheological fluids for the Corvette. The company is also very active in smart suspension systems.

Hercules is advanced in the area of embedded optical fiber sensing systems. The company is forging ahead with plans for demonstrating the manufacturability of embedded optical fiber sensors in advanced composite materials. Hercules is very well positioned to take a step forward in the area of smart materials and structures. The established leadership of the company in the advanced composite materials market gives it a very distinct edge in capitalizing on the impending revolution in smart materials and structures technologies.

The research and development program of Hercules in the area of electro-rheological fluids and related products is well respected by the competition. However, the company has considerably reduced its efforts in this area during the last two years.

Hughes is assessing the benefits and technological impediments of implementing smart skins for the U.S. Airforce.

Imperial Chemical Industries is very active in the chemistry of electro-rheological fluids; ICI is looking at large commercial markets.

Kawasaki Heavy Industries, Japan's number two defense contractor is very active in all aspects of smart technologies; they hope to establish leadership in the area of smart weapons systems by 1995.

Lockheed is focussing on stealth technology based on special conductive plastics for smart skins that would absorb and dissipate radar signals and emit signals to baffle enemy receivers. The company is also exploring concepts based on embedded electro-rheological fluids and piezo-ceramic actuators in order to develop viable smart structures.

Lord is probably yielding in the area of electro-rheological fluids development in the wake of dry fluids developed at the University of Michigan and in Britain. The company is becoming more applications oriented and evaluating several actuating and sensing technologies. Lord has reorganized and substantially increased the size of its smart materials and structures group, and has very good potential for success in this field.

McDonnell Douglas is very active in the area of embedded fiber optic sensors for enhanced manufacturing quality and productivity of commercial and military systems. McDonnell Douglas could have viable smart skin prototypes on the leading edges of F-15 wings by 1992. At this time, the company is leaning towards piezo-electric actuation technologies.

Messerschmitt Bolkow Blohm is extremely active in the areas of fiber optics and piezo-electric materials. It is also carefully evaluating all areas of smart rials and structures technologies. The company is very well positioned to capitalize on the impending revolution in smart materials. Its FORS (Faser-Optischer-Riss-Sensor) fiber optic crack detections system is used on metal surfaces and embedded in glass and carbon fiber composites; FORS is a prime example of the leadership of this company in fiber optic sensing systems. Messerschmitt Bolkow Blohm is currently developing a complete fiber optic nervous system based on FORS technology for in-flight surveillance of the composite structures in the Airbus aeroplanes.

Mitsubishi Electric is developing semiconductor chips, sensor technologies, and superlight superstrong materials for smart weapon systems and various commercial applications.

Mitsubishi Heavy Industries is exploring every aspect of smart materials and structures technologies including semiconductor chips, various sensor technologies and super composite materials. The company is very active in the field of electro-rheological fluids; it is also evaluating memory metals and piezoelectric materials. Mitsubishi is positioning itself to continue its dominance as the premier defense contractor in Japan; they hope to make windfall profits by exporting smart weapon systems to the U.S. once the 1976 ban on military exports is relaxed—possibly by 1995.

NEC is very active in the area of semiconductor chips and sensor technologies that would be crucial to the evolution of smart materials and structures.

Raytheon is actively trying to establish leadership in the areas of advanced avionics and sensors technologies. The conformable array radar serves as a good testbed for the company to demonstrate its significant potential.

Rockwell is assessing the benefits and technological impediments of implementing smart skins for the U.S. Airforce.

Sikorsky is active in sensor technologies associated with smart skins.

Textron Aerostructures is focused on embedded fiber optics for damage assessment. The technology will be potentially employed on Lockheed and Rockwell systems.

Toshiba is focussing its efforts in the area of sensor technologies for smart weapon systems.

Toyota is actively researching all aspects of smart materials and structures technologies in order to develop superior suspension systems, drive train components, anti-lock braking systems and other devices. It is believed that Toyota will be the first company in the commercial sector to extensively exploit smart materials and structures technologies.

Tremac SA de C.U. Queretaro is focussing efforts on brakes, steering, and drive train elements that incorporate electro-rheological fluids technologies. The company has very close ties with the University of Michigan, which has licensed these technologies for the company's use.

United Technologies Corporation, Chemical Systems Division, is developing smart skins that would monitor the integrity of casings around big rockets and simplify prelaunch checks for leaks before space shots.

United Technologies Corporation, Defense and Space Systems Division, is concentrating on smart composites for stealth applications. This classified research and development program is supported by the U.S. Air Force.

The Japanese Effort

The Ministry of International Trade and Industry (MITI) in Japan is actively coordinating several programs in the areas of smart materials and structures under several innocuous headings such as electronics and superlight, superstrong materials. There is tremendous pressure on the Government from major defense contractors such as Mitsubishi Heavy Industries, Kawasaki and Ishikawajima-Harima to relax the 1976 guidelines banning the export of military hardware. Electronic components and sensor technologies are already bypassing this ban, and it is anticipated that the restrictions will be eliminated by 1995.

MITI is coordinating programs that will permit major Japanese defense contractors to export smart weapons systems to the U.S. Whereas Mitsubishi Heavy Industries, Kawasaki, and Ishikawajima-Harima are concentrating on the broader issues of integrating various technologies for smart weapon systems; NEC, Toshiba, and Mitsubishi Electric are focussing on electronics, sensor technologies and superlight-superstrong materials. Most of these companies are also positioning themselves for joint ventures with major U.S. defense contractors.

The Japanese are very confident that they can effectively compete with the established superstars of the U.S. defense industry. The Japanese would seek economies of scale to reduce the cost of their own military hardware and to increase their export market (which would be initially very small). At the same time they would seek to increase research and development expenditures to encourage the introduction of smart technologies in the commercial, automotive and advanced manufacturing sectors, which would render the playing field wide open. Defense procurement will no longer dominate the development of high technology products, and the market would be driven by commercial needs and not necessarily by military considerations. The primary Japanese edge will be based on the fact that electronics and materials technology breakthroughs on the commercial front can easily be applied to defense systems.

It is believed that the legendary Japanese attention to detail, quality control, and reliability of manufactured products give major Japanese companies a distinct edge in the impending revolution in smart materials and structures technologies.

Japanese companies will also establish themselves as reliable defense contractors. Even today, the majority of gallium arsenide chips used in missiles and radar systems in the U.S. are provided by the Japanese. When manufactured in Japan, electronic subsystems used in aircraft or missile control have much higher reliability when compared to identical systems built under license in the U.S. It has been reported that there is a growing perception within the Pentagon that Japan can contribute more advanced defense technology than any other U.S. ally, especially when it comes to smart weapons systems.

9

RESEARCH ACTIVITIES

What the Government, Universities and Major Consultants Are Doing in the Area of Smart Materials and Structures

The race is already on, and major Fortune 500 companies, government agencies, universities, research centers, publishing houses, and consulting companies have already started positioning themselves to reap the windfall profits that will start accruing with the onset of the revolution in smart materials and structures. Some of the key indicators highlighted in the subsequent pages will help one appreciate what this scramble is all about and project the shape and size of things to come.

At least three workshops have been organized during the last twelve months in the area of smart materials and structures. These workshops were organized by NASA, ARO and a private publishing company.

At least four technical reports have been marketed worldwide by private publishing/consulting companies during the last twelve months in the area of smart materials and structures.

At least seven major publishers are in the process of publishing books in this area and at least four publishing companies are in the process of launching new international journals focused on the area of smart materials and structures.

Two major universities have established centers of excellence focused on smart materials and structures during the last twelve months: Michigan State University—Intelligent Materials and Structures Laboratory; and Virginia Polytechnic Institute and State University—Smart Materials and Structures Center.

Several other universities have initiated very focused programs in the area of smart materials and structures.

- Byelorussian SSR Academy of Science
- Cranfield Institute of Technology
- Draper Laboratories
- Georgia Institute of Technology
- Massachusetts Institute of Technology

- Pennsylvania State University
- Stanford University
- State University of New York
- Waseda University
- Ukrainian SSR Academy of Science
- University of Bristol
- University of Liverpool
- University of Sheffield
- University of Tokyo

The U.S. Army Research Office has identified the area of smart materials and structures as one of its top priorities, and its University Research Initiative solicitation released in January 1989 specified key smart technologies in its request for proposals.

The U.S. Department of Defense Small Business Innovative Research Award solicitations, with a deadline of January 1989, identified at least 20 projects focused on smart materials and structures that are relevant to the Army, Navy, Airforce and the Strategic Defense Initiative Office.

The U.S. Airforce has identified the area of smart materials and structures as one of its four top-priority areas to pursue in the 21st century in its Project-Forecast 2.

The U.S. Navy has started supporting the area of smart materials and structures through its Presidential Young Investigator Program.

The U.S. Defense Advanced Research Projects Agency is currently assessing various smart materials and structures technologies at the highest level in order to determine the nature and scope of a major initiative that may be launched in fiscal year 1990.

At least 100 different magazines, newspapers and newsletters have covered the area of smart materials and structures during the last twelve months. These sources include *Business Week*, the *New York Times*, and *Technology Review*.

The first major conference devoted to fiber optic smart structures and skins was held in conjunction with SPIE's OE/FIBER LASER Symposium in Boston in September 1988. The success of the 1988 conference led to the scheduling of a second conference on the subject in Boston in September 1989.

This section of the intelligence report documents the principal research groups actively working on the development of smart materials featuring electro-rheological fluids, shape memory alloys, piezo-electric polymers and embedded fiber optic sensing systems. This section highlights possibilities for collaborative ventures, licensing opportunities and a comprehensive listing of potential consultants in the area.

Research Group: Michigan State University, Intelligent Materials and Structures Laboratory.

Principal Area of Research: All aspects of smart technologies.

The Intelligent Materials and Structures Laboratory was formally established in 1988 in order to prosecute both fundamental and applied research in the broad field of smart materials and structures. The principal investigators of this research group have been active in the area of smart materials and structures for the last 6 years. An interdisciplinary team has been established with diverse backgrounds in mechanical engineering, applied mathematics and mechanics, electro-magneticism, optics, physical chemistry, materials science and rheology in order to address the research issues in this embryonic field. Programs are focused on memory-metals, electro-rheological fluids, piezo-electric materials and fiber optic sensors.

More than twenty scientists and students are involved in this effort that is funded by the Defense Advanced Research Projects Agency, U.S. Army Research Office Innovative Research Programs, National Science Foundation, and the State of Michigan's Department of Commerce, Research Excellence and Economic Development Fund.

Key Researchers: Dr. Mukesh V. Gandhi, Ph.D., 1984, University of Michigan; Dr. Brian S. Thompson, Ph.D., 1976, University of Dundee; Dr. Paul Hunter, Ph.D., 1969, University of Manchester; Dr. Carl Foiles, Ph.D., 1964, University of Illinois; Dr. K. Mukerjee, Ph.D., 1965, University of New York.

Modes of Interaction: The research group is open to licensing of technology, receipt of research contracts from federal and industrial sources, and the principal researchers are open to establishing consulting agreements with industrial partners.

Contact: Professor M. V. Gandhi, Intelligent Materials and Structures Laboratory, Michigan State University, East Lansing, MI 48824-1326. Phone: (517) 355-1744.

Research Group: Electro-Rheology Research Syndicate plc.

Principal Area of Research: Electro-rheological fluids.

This British industrial/university research consortium has the following membership: Air-Log plc, Automotive Products, British Aerospace plc, British Hydromechanics Research Association, British Technology Group, Cranfield Institute of Technology, Castrol Ltd., Dowty Rotol Ltd., GEC plc, and the University of Sheffield. The Universities of Bristol, Liverpool, and Cambridge University are funded by the syndicate to undertake basic studies in ER fluids.

Key Researchers: These individuals reside in the British Universities documented on the adjacent pages and the various industrial partners.

Modes of Interaction: The research group is open to licensing of technology, receipt of research contracts from federal and industrial sources, and the principal researchers are open to establishing consulting agreements with industrial partners.

Contact: Arthur Garish, 01-251-0095. The Electro-Rheology Syndicate Ltd., 101–108 Britannia Walk, London, N1 7LU, U.K.

Research Group: British Technology Group.

Principal Area of Research: Electro-rheological fluids.

This group is responsible for licensing patents resulting from research sponsored by the British Ministry of Defense and the Electro-Rheology Research Syndicate. The group has a large number of patents on ER fluid formulations, manufacturing technologies and equipment, and fluid devices. Beneficiaries of licensing agreements include American Cyanamid, Castrol, and Imperial Chemical Industries.

Key Researchers: These individuals reside in the British Universities documented on the adjacent pages and the various industrial partners.

Modes of Interaction: The research group is open to licensing of technology, receipt of research contracts from federal and industrial sources, and the principal researchers are open to establishing consulting agreements with industrial partners.

Contact: British Technology Group, 101 Newington Causeway, London, SE1 6BU.

Research Group: ER Fluid Developments plc.

Principal Area of Research: Electro-rheological fluids and devices.

The group has ownership of a number of European patents on ER fluids which have been the result of the leadership of J. Stangroom. The company works closely with clients to develop goal-oriented prototypes.

Key Researchers: Dr. J. Stangroom, Ph.D. 1961, University of Sheffield.

Modes of Interaction: The research group is open to the licensing of ER technology and joint product development. The key researchers are open to establishing consulting agreements.

Contact: British Technology Group, 101 Newington Causeway, London, SE1 CBU, U.K. for licensing details. ER Fluid Developments Ltd., Vincent Works, Brough, Bradwell, Sheffield, S30 2HG, Dr. J. Stangroom. Phone: 0433-21449.

Research Group: Virginia Polytechnic Institute and State University; Smart Materials and Structures Laboratory.

Principal Area of Research: Smart Composite Structures featuring Memory Metals.

The laboratory was founded in 1988 and the activities are focused on the development of smart structures featuring shape memory alloys. More than twelve scientists and students are involved in this effort which is partially funded by the Virginia Center for Innovative Technology, the U.S. Office of Naval Research, and U.S. Nitinol, Saratoga, California. The group has applied for a broad patent describing the invention of smart materials featuring shape memory alloys.

Key Researchers: Dr. Craig A. Rogers, Ph.D. 1987, Virginia Polytechnic Institute and State University; Dr. Harold H. Robertshaw, Ph.D. 1971, University of Virginia.

Modes of Interaction: The research group is open to licensing of technology, receipt of research contracts with federal and industrial sources, and the principal members are open to establishing consulting agreements with industrial partners.

Contact: Professor C. A. Rogers, Smart Materials and Structures Laboratory, Department of Mechanical Engineering, Virginia Polytechnic Institute and State University, Blacksburg, VA 24061. Phone: (703) 961-7194.

Research Group: Lord Corporation; Thomas Lord Research Center.

Principal Area of Interest: Electro-rheological fluids and devices.

The electro-rheological fluids group at Lord, which is directed by T. Duclos and D. Carlson, comprises eight scientific staff who have been working in ER fluids and devices for the past five years. The company has focused the efforts of the group on ER fluid devices for automotive applications. The group has been awarded several patents, including at least two patents on engine mounts, and in 1987 it submitted a patent application for an anhydrous fluid.

Key Researchers: Dr. Theodore G. Duclos, Ph.D. 1984, Duke University; Dr. John P. Coulter, Ph.D. 1987, University of Delaware.

Modes of Interaction: The research group is *not* willing to license its patents, however, it is keen to collaborate with industrial partners to develop new products.

Contact: Theodore G. Duclos, ER Fluids Program Manager, Thomas Lord Research Center, 405 Gregson Drive, P.O. Box 8225, Cary, NC 27511. Phone: (919) 469-3443.

Research Group: University of Michigan, Department of Materials Science and Engineering.

Principal Research Group: Field-dependent fluids.

The interdisciplinary research group includes at least ten doctoral and post doctoral students and several faculty members. The group, under the leadership of F. Filisko, has been responsible for one of the major break throughs in ER fluid technology through their patent on an anhydrous field-dependent fluid that was awarded in May 1988. Tremac owns exclusive rights to this anhydrous fluid for brake, steering and drive-train applications. Several other automotive applications have also been licensed. Current funding sources include U.S. materials manufacturers and several end-users; total funding level is estimated to be around $1 million per year. Major Japanese companies are carefully monitoring the activities of this premier research group.

Key Researchers: Dr. F. Filisko, Ph.D. 1969, Case Western; Dr. A. Wineman, Ph.D. 1963, Brown University.

Modes of Interaction: The research group is open to licensing technology for limited domains of applications, such as biomedical applications.

Contact: Robert F. Gavin, Intellectual Properties Office and Intellectual Property Counsel, University of Michigan, Ann Arbor, Michigan 48109. Phone: (313) 763-0614.

Research Group: Virginia Polytechnic Institute and State University, Fiber and Electro-Optics Research Center.

Principal Areas of Research: Optical Fiber Sensing Systems

The laboratory, founded in 1980, focuses activities on the development of optical fiber sensors and optical signal processing strategies for determining strain, temperature resin cure, acoustic emission, and impact and battlefield damage. More than 15 scientists and graduate students are involved in these efforts; currently the laboratory has more than $1 million U.S. dollars in funding from both the federal government and industry. Funding sources include NASA Langley, Hercules Aerospace, Virginia Center for Innovative Technology, Lockheed, General Dynamics, McDonnell Douglas, and Martin Marietta.

Key Researchers: Dr. R. Claus, Ph.D., 1977, John Hopkins.

Modes of Interaction: The research group is open to licensing of technology, receipt of research contracts with federal and industrial sources, and the principal members are open to establishing consulting agreements with industrial partners.

Contact: Professor R. Claus, Fiber and Electro-Optics Research Center, Virginia Polytechnic Institute and State University, Blacksbury, VA 24061. Phone: (703) 961-7203.

Research Group: Georgia Institute of Technology, School of Aerospace Engineering.

Principal Areas of Research: Piezo-Ceramic and PVDF Film Sensors and Actuators, and Associated Controls Strategies for Smart Structures.

The activities of this research group are focused on the development of viable distributed controls strategies for smart materials and structures by employing piezo-ceramic and PVDF films. This work is supported by the U.S. Army Research Office through its Center of Excellence on Rotorcraft Systems.

Key Researchers: Dr. S. Hanagud, Ph.D.

Research Group: Pennsylvania State University, Materials Research Laboratory.

Principal Area of Research: Electro-Ceramic Actuators and Sensors, and applications to opto-electronic systems. Professor Newnham, who directs this Laboratory, has been active in the field of electro-ceramics for over thirty years. Current work is focused on nano-composites.

Contact: Dr. R. E. Newnham, Pennsylvania State University, Materials Research Laboratory, University Park, PA 16802. Phone (814) 865-1612.

Research Group: Massachusetts Institute of Technology, Space Systems Laboratory.

Principal Area of Research: Embedded piezo-electric actuators and sensors and associated controls strategies for smart structures for space applications. The group has substantial funding from NASA to develop smart trusses and other smart structural systems. The group has diverse strengths in mathematical, analytical and experimental areas and has a good balance between research and development activities.

Contact: Dr. E. F. Crawley, Aeronautical and Astronautical Engineering Department, MIT, Cambridge, MA.

Research Group: Charles Stark Draper Laboratory.

Principal Area of Research: Embedded piezo-electric actuators and related control issues and strategies for smart structures. Research has been mainly academic in nature and focused on beam-like structures.

Contact: Dr. J. E. Hubbard, C. S. Draper Laboratory, 555 Technology Square, Cambridge, MA 02139.

Research Group: Industrial Technology Institute, Advanced Manufacturing Technologies Laboratory.

Principal Area of Interest: Smart fixtures and robotic grippers for automated manufacturing environments.

The smart fixtures and grippers group at the Industrial Technology Institute is directed by Dr. H. Stoll for the last five years. The group comprises a scientific and technical staff of seven professional researchers and engineers, and the group has focused efforts on deploying electro-rheological fluids and memory metal alloys and other innovative technologies for the development of a new generation of fixtures and grippers. The very active group has several patent applications pending in this area.

Key Researchers: Dr. H. Stoll, Ph.D. 1967, University of Illinois; P. Grippo, M.S. 1987, University of Michigan.

Modes of Interaction: The research group is open to licensing possibilities and collaborations with industrial partners to develop new technologies and products.

Contact: Paul M. Grippo, Smart Fixtures and Grippers Program Coordinator, Industrial Technology Institute, Hubbard Drive, Ann Arbor, MI 48109. Phone: (313) 769-4000.

Research Group: Hy-Tech International, Okemos, MI.

Principal Areas of Research: All aspects of smart materials and structures technologies, technology transfer issues, reports and seminars.

Hy-Tech International is a full-service consulting company in the area of advanced materials offering clients multi-dimensional capabilities in technology-transfer, marketing, design and manufacture. The company, founded in 1984, has subsequently provided services to several Fortune 500 companies. The company comprises a diverse group of twenty professionals that includes research scientists, business management consultants and economists. Since the professional staff at Hy-Tech International serves as consultants to various government agencies, Fortune 500 companies and major high-technology-oriented companies worldwide, this report provides only a qualitative technological profile of the activities of these companies. The professional staff at Hy-Tech International are at your disposal to provide technical and management services that can catalyse the evolution of a competitive edge for your company in this rapidly evolving field.

Key Researchers: Dr. M. V. Gandhi, Ph.D., 1984, University of Michigan; Dr. B. S. Thompson, Ph.D., 1976, University of Dundee.

Modes of Interaction: The company is open to licensing possibilities and collaborations with industrial partners to develop new technologies and products. The company also provides management and technical services in all aspects of smart materials and structures technologies, technology transfer issues, reports, and workshops.

Contact: Dr. M. V. Gandhi or Dr. B. S. Thompson, Hy-Tech International, 2112 White Owl Way, Okemos, MI 48864. Phone: (517) 355-1744, (517) 355-2179.

Other Players

Organization	Key Researchers	Key Specialty
Cranfield Institute of Technology (U.K.)	Dr. H. Block	Dry-fluids
Sheffield University (U.K.)	Dr. W. A. Bullough	ER-based devices
Liverpool University (U.K.)	Dr. J. L. Sproston Dr. P. Stanway	ER-based devices and vibration control
Bristol University (U.K.)	Dr. J. Goodwin	Constitutive modeling of ER fluids
Waseda University (Japan)	Dr. Y. Miwa Dr. H. Uejima	Shape memory devices for robotic applications ER fluids
BSSR Academy of Sciences (USSR)	Dr. Z. P. Shulman	ER-based devices
NASA/Langley	Dr. J. Heyman Dr. R. Rogowski: (804) 865-3036	Fiber optics
McDonnell Douglas Astronautics Company	Dr. E. Udd: (714) 896-3311 Ext: 41691 Dr. H. Smith	Fiber optics, smart skins

Organization	Key Researchers	Key Specialty
Hercules	Dr. D. Simpson Dr. S. Ahmad Dr. W. Spillman (802) 877-2911 Ext: 2539	Fiber optics, smart skins
MBB (West Germany)	Dr. Brandt	Piezo materials fiber optics
University of Minnesota	Dr. James	Memory Metals
State University of New York	Dr. D. Inman	Piezo-electrics
Ukrainian SSR Academy of Science (USSR)	Dr. U. Deinega	ER fluids
University of Kentucky	Dr. Wary	Piezo-electrics
Boeing	Dr. Bates	Smart technologies
University of Toronto Ontario, Canada	Dr. Tennyson Dr. Measures	Fiber optics, leading-edge of wings
U.S. Nitinol Saratoga, CA	Dr. Chuck Hovey (408) 741-5563	Shape memory alloys
Textron Aerostructures Nashville, TN	Dr. L. Wood Dr. M. Praire	Smart skins, fiber optics
Memory Metals, Inc. Norwalk, CT	Dr. Schetky (203) 853-9777	Shape memory alloys
Raychem Menlo Park, CA	Dr. Bryant (415) 361-4394	Shape memory alloys
Fiat (Italy)	Dr. G. Manfre	Memory-metals, ER fluids
Ford Motor Company	Dr. G. Chaundy	Smart materials
General Motors	Dr. G. Smith (313) 986-0592	ER fluids
Lockheed, Missiles and Space Company	Dr. L. Weisstein (408) 756-4653	ER fluids

APPENDIX A: SIGNIFICANT PATENTS

1. Electro-Rheological Fluids

4767551

Tunable electro-rheological fluid mount for vehicle—has valves located upon opposite sides of central mount section through which chamber connecting passages extend.

Patent Assignee: LORD CORP

Inventor: DUCLOS, T.; HODGSON, D.; CARLSON, J.

Date: 880302

Designated States (Regional): DE; FR; GB; IT; SE

4745375

Inertia type fluid mount using electro-rheological fluid for vehicle—has energisation of electrode type of valve associated with set chamber producing HV field and fluid solidification.

Patent Assignee: LORD CORP

Inventors: DUCLOS, T. G.; HODGSON, D. A.; CARLSON, J. D.

Date: 880119

4736494

Flexible shaft—comprises sleeve containing electro-rheological fluid with viscosity controlled by external electromagnets.

Patent Assignee: KAUN POLY

Inventors: MEDZHAUSHE, A. A.; POTSYUS, Z. Y. U.; RINKYAVICH, B. B.

Date: 870607

4660922

Electro-hydraulic transducer for hydraulic regulating valve—has insulated tube with concentric bolt, both with conducting surfaces and electro-rheologic fluid passing through gap between tube and bolt.

Patent Assignee: BOSCH R GMBH

Inventor: WEIGLE, D.

Designated States (National): JP; US *(Regional):* AT; BE; CH; DE; FR; GB; IT; LU; NL; SE

4659798

Compression fuel injector for IC engine—utilises electro-rheological fluid to control amount of fuel to be delivered to engine combustion chamber.

Patent Assignee: ELLIOT, G. D.

Inventor: ELLIOT, G. D.

Date: 871020

TWO LETTER COUNTRY CODES

AT: Austria	**BE:** Belgium	**CH:** China	**DE:** Denmark	**FR:** France
GB: Great Britain	**IT:** Italy	**LI:** Lichtenstein	**LU:** Luxembourg	**NE:** Netherlands

4655488

Electro-rheological fluid—contg. insulating particles giving change in apparent viscosity in an electrical field; PRINT CIRCUIT BOARD PCB TEST.

Patent Assignee: GEC AVIONICS LTD

Inventor: BYATT, D. W. G.

Date: 871104

4633215

Variable camshaft drive for IC engine—has space in cam filled with electro-rheological fluid, controlled electrically to couple or decouple drive.

Patent Assignee: FORD MOTOR CO LTD

Inventor: TSOIHEMA, T.

Date: 871014

4530971

Turbine flow rate regulator has body made with spacer forming cavity with strapping, filled with electro-rheological suspension.

Patent Assignee: MOSCOW BAUMAN TECH COLL

Inventor: DEVISILOV, V. A.

Date: 861115

4493245

Penile implant using electro-rheological fluid hardener has electrodes supplied with power with potential difference making fluid more viscous and so causing erection.

Date: 870512

4394825

Viscous damper for active vibration isolation system has servo-valve using electro-rheological fluid to couple load supporting actuator to accumulator.

Patent Assignee: BARRY WRIGHT CORP

Inventor: SCHUBERT, D. W.

Date: 870225

4342054

Material friction test device has vessel with float and liquid to compensate weight of support.

Patent Assignee: LENGD KALININ POLY

Inventor: EGOROV, V. I.; LIVOTOV, P. L.; ORLENKO, E. O.

Date: 860507

4321548

Locality relief modelling device has insulated electrodes positioned on substrate and connected to power source.

Patent Assignee: KUIB GIDROPROEKT

Inventors: BORSKII, O. B.; RITERMAN, E. S.; ELISEEV, P. I.

Date: 860423

4274807

Rotating shaft angular vibro-displacements meter uses electro-rheological fluid as inertial mass and regulation system to control viscosity of fluid to optimum sensitivity.

Patent Assignee: AS BELO MACH RELIAB

Inventors: BERESTENEV, O. V.; ISHIN, N. N.; BAINYUK, V. L.

Date: 860330

4210675

Acoustic antenna beam steering for scanning purposes modifies transmission characteristics of acoustic cell containing electro-rheological fluid using electric field.

Patent Assignee: GENERAL ELECTRIC CO PLC

Inventors: COLLINS, R. J. W.; ELLIS, A. B. E.

Date: 860813

4202983

Electro-rheological fluid contg. liq. continuous phase and disperse phase functioning when anhydrous; TRANSFORMER OIL.

Patent Assignee: NATIONAL RES DEV CORP

Inventors: BLOCK, H.; KELLY, J. P.

Designated States (Regional): AT; BE; CH; DE; FR; GB; IT; LI; NL; SE

4150841

Pipeline gas flow remote blocking device has electro-rheological suspension fed into barrier of electrodes placed across the pipe bore.

Patent Assignee: HEAT MASS TRANSFER

Inventors: SHULMAN, Z. P.; RAGOTNER, M. M.; GORODKIN, R. G.

4098326

Electrically controlled progressive frictional torque transmitter controls degree of frictional engagement by application of variable potential to electro-rheological fluid within clutch.

Patent Assignee: ER FLUID DEV LTD

Inventor: STANGROOM, J. E.

Designated States (Regional): AT; BE; CH; DE; FR; BG; IT; LI; LU; NL; SE

4089659

Weight-lifting mechanisms test loading device uses commutator to control connection of electrodes and phase of electro-rheological fluid in tank.

Patent Assignee: PROKHODA, O. K.

Inventors: PROKHODA, O. K.; SPIRIN, V. G.; FASTOVERTS, P.

Date: 850915

4073000

Servo device with pressure indicator for vehicle brakes has double or stepped piston with piezo-electric pressure sensor and electro-rheological valve.

3998618

Programmable bed-of-nails test access jigs with electro-rheological fluid actuation, preferably using silicone oil-based fluid containing a suspension of polymethyl methacrylate.

Patent Assignee: GENERAL ELECTRIC CO PLC

Inventor: COLLINS, R. J. W.

Date: 860102

Designated States (Regional): AT; BE; CH; DE; FR; IT; LI; LU; NL; SE

3948015

Fluid pressure oscillation generator has closed converter chamber and separator in form of elastic membrane.

Patent Assignee: HEAT MASS TRANSFER

Inventors: SHULMAN, Z. P.; BLOKH, G. M.; GORKUN, V. I.

Date: 850423

3947963

Variable moment of inertia device has conical electrodes secured to base in line with rotary shaft.

Patent Assignee: MICRONCHENKO, V. I.

Inventors: MIRONCHENKO, V. I.; PRONIKOV, A. S.

Date: 850423

3941767

Downhole angle finder with transducer uses eyepiece adjusting on nonius scales, colour marked float, and transducer located base.

Patent Assignee: KAZA MINERAL RAW MA

Inventors: LOSKUTOV, G. I.; ESTERLE, O. V.; PODGORNYI, P. P.

Date: 850415

3927679

Viscous substance dispenser has rotary cylinder with piston to which elastic membrane is attached.

Patent Assignee: LITH FOOD IND DES

Inventors: SHLAMAS, M. A.; STRAVINSKA, Y. U. K.; SHLAMAS, V. A.

Date: 850323

3864974

Electro-rheological fluids comprising a solid particulate substance in a liquid halogenated aromatic hydrophobic vehicle.

Patent Assignee: NATIONAL RES DEV CORP

Inventors: STANGROOM, J. E.; HARNESS, I.

Date: 850807

Designated States (Regional): DE; FR; GB; IT

3788436

Hydrodynamic transmission for centrifuge drive uses an electro-rheological suspension linked with a variable power source to alter fluid viscosity over wide range.

Patent Assignee: MOSCOW CHEM EQUIP. INST.

Inventors: KOLTUNOV, V. V.; SOTSKOVA, I. G.; SOTSKOV, Y. U. N.

Date: 841030

3765192

System elastic characteristics regulating process using elastic-viscous electro-rheological suspension with frequency-regulated tension.

Patent Assignee: ANASKIN, I. F.

Inventors: ANASKIN, I. F.; GLEB, V. K.; KOROBKO, E. V.

Date: 840930

3747730

Electro-rheological fluid actuator, e.g., for vibrator includes controller connected to electrode assemblies to apply electric field across flow path to vary valve flow resistance; VISCOSITY.

Patent Assignee: UK SEC FOR DEFENCE

Inventor: STANGROOM, J. E.

Date: 850207

Date: 830801

3728360

Laboratory animal respiration parameter meter has conical part which is electrically non-conductive with insulated fastening electrodes on inner surface.

Patent Assignee: KUIB HYGIENE RES

Inventors: PASHKOVA, G. A.; BORSKII, O. B.; KLIMOV, V. S.

Date: 840807

3722209

Manipulator module for assembly work has flexible non-tensile tubes, resilient inserts and electrodes with electro-rheological fluid filled spaces.

Patent Assignee: AS BELO CYBERNETICS

Inventors: BEREZOVIK, V. V.; LUKYANOVIC, A. V.; LUKYANOVIC, D. V.

Date: 840723

3683948

Industrial robot gripper jaw comprises a fixed and moving section interconnected by flexible elements.

Patent Assignee: BYELORUSSIAN POLY

Inventors: LUKYANOVIC, A. V.; LUKYANOVIC, D. V.; MYSHKOVSKI, V. V.

Date: 840523

3672153

Azimuth and zenith angle finder for boreholes uses oxide coated electrodes on chamber filled with electro-rheological fluid to arrest float.

Patent Assignee: KAZA MINERAL RAW MA

Inventors: LOSKUTOV, G. I.; PODGORNYI, P. P.; ESTERLE, O. V.

Date: 840507

365688

Eye ball attachment device has electrodes on end face covered with electro-rheological suspension.

Patent Assignee: KUIB HYGIENE INST

Inventors: RITERMAN, E. S.; PASHKOVA, G. A.; BORSKII, O. B.

Date: 840330

3548149

Vibrating assembly platform for peg in bush assembly has platform supported on fluid bed for improved performance and wider range of applications.

Patent Assignee: SEVAST INSTRM INST

Inventors: YAKHIMOVIC, V. A.; KHANSHCH, Y. U. A.; LEBEDEVOSK, M. S.

Date: 831130

3541063

Electro-conducting surface unevenness test device has electrode with electro-rheological suspension on contact surface whose separation force determines unevenness.

Patent Assignee: GORODKIN, R. G.

Inventors: GORODKIN, R. G.; GELIKMAN, B. Y. U.; ZALATA, V. N.

Date: 830423

3508720

Electro-rheological machine to be used as fluid pump has ends of path communicating with inlet and outlet ports that may be shielded by each other by fixed partition or wiper.

Patent Assignee: UK SEC FOR DEFENCE; NATIONAL RES DEV CORP

Inventor: STANGROOM, J. E.

Date: 840620

3504657

Hydraulic manipulator with control and activating mechanism has separating cylinder filled with electro-rheological fluid for locking actuating mechanism.

Patent Assignee: DOROKHOV, V. P.

Inventors: DOROKHOV, V. P.; GORBUNOV, V. A.

Date: 830923

3480026

Rotational viscosity meter has cylindrical electromagnetic rigidly connected to push rod linked with drive.

Patent Assignee: MOSC FOOD IND TECHN

Inventors: MARSHALKIN, G. A.; GORYACHEVA, G. N.; SHLAMAS, M. A.

Date: 830815

3421886

Endless belt type pressing set has sources of magnetic or electric field inside loops of endless hose-type belts bounding pressure chamber.

Patent Assignee: AS BELO METAL POLYM

Inventors: AZBUKIN, K. S.; YUNITSKII, A. E.; BROZHAT, A. A.

Date: 830515

3322676

Automatic manipulator gripper has clamping chamber with elastic metering chamber filled with fluid and connected to pressure relay.

Patent Assignee: SHVARTSMAN, A. Z.

Inventor: SHVARTSMAN, A. Z.

Date: 830115

3307076

Electrically variable linkage joint for industrial robot or prosthesis utilises properties of electro-rheological fluid to provide fast response linkage which can be interfaced with electronic control system.

Patent Assignee: UK SEC FOR DEFENCE; NATIONAL RES DEV CORP

Inventor: STANGROOM, J. E.

Date: 831102

3286012

Device for stabilising electric motor speed has regulating contacts immersed in electro-rheological liquid with viscosity altered by control voltage.

Patent Assignee: BUSHUEV, V. A.

Inventors: BUSHUEV, V. A.; GRIGOREV, S. I.; KOMARKOVSK, L. I.

Date: 821215

3266858

Crystal growing hydraulic drive speed control uses electro-rheological throttles in pockets of insulating material rings on rod.

Patent Assignee: KAUN POLY

Inventors: MALISHAUSK, M. A.; RAGULSKIS, K. M.; VOLNGYAVIC, A. A.

Date: 821030

3237396

Pipe bending method uses a flexible electrode running through an electro-rheological fluid inside the pipe to apply constant voltage on electrode.

Patent Assignee: AS BELO DES BUR TES

Inventors: KURDYA, V. F.; MARUTYAN, M. P.

3219146

Quick-release pipe joint has electromagnet and flexible ring filled with electro-rheological fluid for reliability.

Patent Assignee: KAUN FOOD IND DES

Inventors: SHLAMAS, M. A.; STRAVINSKA, Y. U. K.; SHLAMAS, V. A.

Date: 820917

3137656

Electrically operated linear variable viscosity damper eliminates risk of fluid cavitation and air induction by using atmospheric pressure sensors coupled to controller.

Date: 830629

3132430

Electro- or magneto-viscous fluid coupling transmission ratio control uses pulse action of field on fluid to increase accuracy and efficiency.

Patent Assignee: BABAEV, V. A.

Inventors: BABAEV, V. A.; TEREKHIN, I. I.

Date: 820817

3111750

Dynamic rotor balancer has sensors to produce signals corresponding to rotor imbalance.

Patent Assignee: KAUN POLY

Inventors: BANSEVICHY, R. Y. U.; VOBOLIS, I. P.; IONUSHAS, R. A.

Date: 820715

3095202

Ultrasonic monitoring detector has sectioned sump and electrodes filled with electro-rheological fluid that alters the viscosity and acoustic resistance.

Patent Assignee: AS UKR HARD MATERIALS

Inventors: DOBROVOLSK, G. G.; TONKAL, S. S.; KAMCHATNY, Y. U. G.

Date: 820523

3094855

Pipe bending mandrell body base is solid with flexible, closed centre section and traction drum drive for compactness and simplicity.

Patent Assignee: TEREKHIN, I. I.

Inventor: TEREKHIN, I. I.

Date: 820523

3091748

Industrial robot hydraulic drive improves positional control of solid wall of flexible chambers through activation of electrical field on electro-rheological fluid.

Patent Assignee: GORKI MACH CONS COM

Inventor: SMIRNOV, N. N.

Date: 820515

3043994

Tube bending mandrel utilises an electro-rheological liquid filler that solidifies in a HV electric field.

Patent Assignee: TEREKHIN, I. I.

Inventor: TEREKHIN, I. I.

Date: 811215

2873450

Shuttering panel for use in mines has a magnetically, electrically, or thermally activated filler to hold panel shape during concrete placing.

Patent Assignee: TEREKHIN, I. I.

Inventor: TEREKHIN, I. I.

Date: 811123

2857670

Variable inertia accumulator unit has electro-rheological fluid between two insulated, electrically conducting discs, which taper inwards to shaft.

Patent Assignee: KAUN POLY

Inventors: BANSEVICH, R. Y. U.; VASNELIS, A. I.; RAGULSKIS, K. M.

Date: 811026

2848912

Ink printing mechanism feeds electro-rheological ink through a tube to a working gap; voltage is applied to stop print operation.

Patent Assignee: AS USSR PHUS CHEM

Inventors: PETRZHIK, G. G.; CHERTKOVA, O. A.; TRAPEZNIKO, A. A.

Date: 811015

2811513

Piezo-electric vibro-drive has hollow ring-shaped damping element filled with variable viscosity, electro-rheological fluid having two electrodes immersed in fluid.

Patent Assignee: KAUN POLY

Inventors: SHTATSAS, L. A. L.; RAGULSKIS, K. M.; KURYLO, R. E.

Date: 810730

2798163

Remote controlled manipulator joint has chamber filled with electro-rheological liquid for controlled damping of moving units via axle mounted discs.

Patent Assignee: GORBUNOV, V. A.

Inventor: GORBUNOV, V. A.

Date: 810722

2794874

Pivot joint with body and moving link has sealed cavity between them filled with electro-rheological fluid to which voltage is supplied.

Patent Assignee: GORBUNOV, V. A.

Inventor: GORBUNOV, V. A.

Date: 810630

2783876

Viscous shear clutch assembly uses electro-rheological fluid as the transmission fluid with differential arrangement of clutch plates that cancels residual drag.

Patent Assignee: UK SEC FOR DEFENCE

Inventor: STANGROOM, J. E.

Date: 820324

613477

Friction transmission with electro-rheological fluid has dielectric disc driving unit and sleeve shaped driven unit each with its own electrodes.

Patent Assignee: KAUN POLY

Inventors: BANSEVICHY, R. Y. U.; BISIGIRSKI, G. V.; RAGULSKIS, K. M.

Date: 810303

2. Shape Memory Alloys

4735210

Linear shape memory treatment for shape memory alloy wire obtained by intermediate coiling, applying surface tension force and electrically heating.

Patent Assignee: FURUKAWA TOKUSHUKIN; FURUKAWA ELECTRIC CO.

Date: 871212

4681528

Device for regulating transfer of heat in heat pipe consists of valve body and valve seat controlled by configuration memory alloy member.

Patent Assignee: VOLVO AB

Inventor: NILSSON, H. O.

Date: 871119

4648844

Low temperature nickel-titanium-chromium shape memory alloy may contain iron to reduce operating temperature.

Patent Assignee: KEIJO KIOKU GOKIN G

Date: 870917

4648843

Low temperature nickel-titanium-iron shape memory alloy can be raised to room temperature operation by adding iron.

Patent Assignee: KEIJO KIOKU GOKIN G

Date: 870917

4634047

Production of shape memory alloy materials—by final shape memory heat treating shape memory nickel-titanium alloy while applying strain after cold work.

Patent Assignee: NIPPON STAINLESS KK

Date: 870903

4586582

Manufacture of titanium-nickel alloy wire by initially hot groove rolling a compacted and sintered mixture of titanium and nickel powders.

Patent Assignee: KANTO DENKA KOGYO

Date: 870721

4525582

Titanium-nickel alloy production by diffusion of titanium-nickel composite.

Patent Assignee: NIPPON SEISEN CO LT

Author Inventor: ISHIBE, H.

Date: 870701

4437014

Production of hollow cored shaped parts from shape memory alloy powders.

Patent Assignee: US SEC OF NAVY

Inventor: GOLDSTEIN, D.

Date: 870106

4304261

Electromechanical magnitude converter for friction motor actuators has transducer comprising Peltier elements bonded between unidirectional shape memory alloys.

Patent Assignee: MITSURBISHI HEAVY IND KK

3964250

Process for conditioning shape memory alloys comprising successive mechanical and temperature constraints and release of constraints in given order.

Patent Assignee: SOURIAU ET CIE

Inventors: RIZZO, G.; DUBERTRET, A.; GUENIN, G.

Date: 851121

3960370

Heat treating shape memory copper-aluminium-nickel-titanium alloy involves quenching from temperature related to beta solid solution transformation temperature.

Patent Assignee: KOBE STEEL KK

Date: 851007

3960369

Heat treatment of copper-zinc-aluminium shape memory alloy by quenching at specified temperature.

Patent Assignee: KOBE STEEL KK

Date: 851007

3823262

Constant temperature maintenance apparatus opens and closes window of electronic equipment using deformation of shape memory alloys.

Patent Assignee: CANON KK

Date: 850511

3812132

Shape memory alloy comprises copper, aluminium, beryllium, zinc and at least one transition metal.

Patent Assignee: SUMITOMO SPEC METAL KK

Date: 850427

3811539

Electronic print circuit board with curved face has shape memory alloys on wafer or portion for making curve face and processing the curing of the face.

Patent Assignee: TOSHIBA KK

Date: 850426

3809519

Suppressing two-way effect in nickel—titanium shape memory alloys including cold working and low temperature annealing steps.

Patent Assignee: RAYCHEM CORP

3800412

Temperature switch has elastomer of bidirectional shape memory alloys that links with conductor and contacts.

Patent Assignee: NISSHIN ELECTRICAL KK

Date: 850415

3792220

Heat sensitive element consists of assembly of shape memory alloy members initially strained at a temperature above or below the transformation point.

Patent Assignee: TOSHIBA KK

3791782

Shape memory alloy fibre filters using, e.g., titanium-nickel alloy, for filtering gases or liquids.

Patent Assignee: DAIDO TOKUSHUKO KK

Date: 850404

3759737

Shape memory alloy comprising nickel, titanium, and vanadium showing constant stress strain behaviour.

Patent Assignee: FAYCHEM CORP

Inventor: QUIN, M. P.

Date: 850319

3747234

Processing beta-phase nickel titanium-based alloys by warm working and warm annealing.

Patent Assignee: RAYCHEM CORP; NIPPON KOKAN KK

Inventors: DUERIG, T.; MELTON, K.

Date: 850305

3663495

Nickel titanium shape memory alloys machined at temperatures above the reverse transformation point.

Patent Assignee: SUMITOMO ELEC IND KK

Date: 841031

3632697

Bonding ceramic workpiece to metal workpiece involves coating ceramic with metal and inserting shape memory alloy as intermediate member.

Patent Assignee: MITSUI ENG & SHIPBUILD

Date: 840926

3619358

Deformable composite material consists of two parts that have different degrees of deformation and execute different movements.

Patent Assignee: BBC AG BROWN BOVERI CIE

Inventors: ALBRECHT, J., DUERIG, T.

Date: 841024

3559370

Forming composite material plastics mould by heating in shape memory alloy mould a reinforcing fibrous material impregnated with matrix resin alloy mould.

Patent Assignee: MITSUBISHI RAYON KK

Date: 840709

3510566

Shape memory alloys with high restoration percentage contain carbon, silicon, manganese, and soluble aluminium plus iron and optional nickel or chromium.

Patent Assignee: SUMITOMO METAL IND KK

Date: 840515

3485712

Titanium-nickel-based shape memory alloys production by melting raw materials in high frequency induction furnace, without contamination by refractor in furnace; HF.

Patent Assignee: CHUO DENKI KOGY KK

Date: 840417

3374585

Composite shape memory alloy is made by sintering mixture of conventional shape memory alloy and a pure metal.

Patent Assignee: HITACHI METAL KK

Date: 831207

3314185

Composite shape memory alloy comprises at least two alloys with different critical temperatures.

Patent Assignee: HITACHI METAL KK

Date: 830928

392460

Alloy for golf club head comprising copper alloyed with zinc, aluminium, antimony or silicon; or nickel alloyed with aluminium or titanium..

Patent Assignee: TAGUCHI, C.

Date: 830421

3040226

Copper alloys showing shape memory effect contain aluminium, beryllium and optional zinc.

Patent Assignee: SUMITOMO SPEC METAL KK

Inventor: SUZUKI, K.

Date: 821014

3033694

Working nickel-titanium-copper shape memory alloys by isothermal extrusion under controlled conditions.

Patent Assignee: BBC AG BROWN BOVERI CIE

Inventors: MERCLER, O.; RICHTER, D.; SCHROEDER, G.

2974361

Shape memory alloys containing nickel, copper and less than stoichiometric amount of titanium.

Patent Assignee: RAYCHEM CORP

Inventor: HARRISON, J. D.

2703466

Modifying transition temperature range of shape memory alloys by deforming, annealing, and then adjusting annealing temperature.

Patent Assignee: US SEC OF NAVY

Inventor: GOLDSTEIN, D.; JONES, R. E.; SERY, R. S.

Date: 810811

2701528

High-homogeneity nickel-titanium shape memory produced by mixing, melting and casting particles.

Patent Assignee: US SEC OF NAVY

Inventors: GOLDSTEIN, D.; HOOVER, S.

2700134

Shape memory alloys of specific transition temperature are produced using prealloyed powders of nickel and titanium.

Patent Assignee: SPECIAL METALS CORP

Inventors: FOUNTAIN, R. W.; BOESCH, W. J.; REICHMAN, S. H.

4769368

Recovery of noble metals using phosphoric extracting agent and monoalkyl-phosphoric acid as accelerating agent.

Patent Assignee: TANAKA KIKINZOKU KOGYO

Date: 880122

4634165

Multi-temperature operation device—laminates several shape memory metals that have different operation temperatures and deforming quantity.

Patent Assignee: TOYOTA JIDOSHA KK

Date: 870903

4600024

Alloy production in plasma—by melting, e.g., copper, aluminium and nickel in plasma then spraying on substrate.

Patent Assignee: TOSHIBA KK

Date: 870731

4426428

Cold trap operating at different temperatures comprises wire mesh filter, by-pass conduit, and valves formed from shape memory metal.

Patent Assignee: DORYOKURO KAKUNENRYO KAI

Date: 870213

4422221

Wire engine for water pump has alternate fluid flow provided by hot-cold air draught to vary shape of alloy wires.

Patent Assignee: OHARE, L.

Inventor: OHARE, L.

Date: 870303

4136270

Consumable electrode for melting and casting metals is coated with metal or alloy.

Patent Assignee: KANTO TOKUSHU SEIKO KK

Date: 860416

4051939

Temperature sensing switch for air conditioning compressor of vehicle has contact mechanism opened and closed by two serially-interconnected shape memory alloy members with different transformation points.

Patent Assignee: DIESEL KIKI KK

Inventors: TAKAHASHI, T.; TAKAHASHI, H.

4036641

Sling line for laying aerial cable on pylon has clamp ring metals of shape memory alloy for holding and fixing cable at equal intervals.

Patent Assignee: NIPPON TELEG & TELEPH

Date: 851224

4022368

Nickel-titanium system alloy with excellent machinability containing sulphur, lead, bismuth, selenium, tellurium and cubic crystal metal(s).

Patent Assignee: DAIDO TOKUSHUKO KK

Date: 851209

3887506

Nickel-titanium alloy production by solidifying molten mixture of nickel and titanium in container containing other metal.

Patent Assignee: MATSUSHITA ELEC IND KK

Date: 850722

3712584

Production of amorphous metallic material by irradiating metal with an electron beam.

Patent Assignee: OSAKA UNIVERSITY

Author (inventor): MORI, H.; FUJITA, H.

3672012

Metal casting with belt mould has belt mould made of metal with shape memory ability to form mould when heated.

Patent Assignee: TEMSKII, S. P.

Inventors: TEMSKII, S. P.; KUZNETSOVA, N. E.

Date:

3645298

Manufacture of shape memory alloy involves mechanically processing assembly of rods of different materials with enclosing member forming billet.

Patent Assignee: MITSUBISHI ELECTRIC CORP

Date: 840515

3638798

Shape memory nickel-titanium alloy is heat treated to produce fine dispersion of intermetallic nickel-titanium compound.

Patent Assignee: FURUKAWA ELECTRIC CO

Date: 840215

3632875

Manufacture of nickel-titanium shape memory alloy involves annealing, cold deforming, and heat treating under specified conditions.

Patent Assignee: FURUKAWA ELECTRIC CO

Date: 840926

3628226

Manufacture of fine copper alloy wire from melt by cold drawing product obtained by directing stream of melt from hole in crucible into coolant on wall of rotating drum.

Patent Assignee: SUMITOMO ELEC IND KK

Date: 840925

3558378

Shape memory titanium-nickel alloy manufacture involves heating lamination of titanium and nickel strips, wires or pipes.

Patent Assignee: SUMITOMO ELEC IND KK

Date: 840705

3558377

Shape memory alloy manufacture by thermal diffusion of laminate of nickel and tin layers.

Patent Assignee: SUMITOMO ELEC IND KK

Date: 840705

3446162

Functional copper alloy member manufacture with improved dimensional accuracy and external appearance by strain-working alloy material of fine-grained structure to give shape memory, super-elasticity etc.

Patent Assignee: SUMITOMO ELEC IND KK

Date: 840302

3425451

Forming dies for sheet metals where male and female dies are formed of shape memory alloy.

Patent Assignee: KANAI H

Date: 840208

1945791

Test for martensitic transformation in alloys with shape memory—includes quenching, bending and measuring deformation redn. temp. for speed and accuracy.

Patent Assignee: KKR METALLOPHYS; KIEV POLY

Authors (inventors): KORCHAK, V. P.; LARIN, V. K.; TITOV, P. V.

Date: 791028

1919546

Beta brass alloy with shape memory—contains copper, zinc and aluminium, with an addition of nickel permitting the alloy to be brazed.

Patent Assignee: BBC AG BROWN BOVERI CIE

Inventors: MELTON, K.; MERCIER, O.

1757649

Temperature responsive actuator system—uses resilient members made of shape memory alloys acting as springs.

Patent Assignee: DELTA MATERIALS RES

Inventors: HART, W. B.; WILLIAMS, R. T.

1685599

Production of solid bodies from copper-zinc-aluminium alloys(s)—by cold compacting the powdered alloy with a beta crystalline structure, and hot extruding the compacted material.

Patent Assignee: LEUVEN RES DEV VZW

Date: 790615

4277870

Copper-base shape memory alloy manufacture by diffusion penetration of zinc into copper alloy.

Patent Assignee: KAGAKU GIJUTSU-CHO KINZ

Date: 860910

4085888

Iron-manganese silicon alloy with improved shape memory property.

Patent Assignee: NIPPON STEEL CORP

Inventors: MORI, T.; SATO, A.; SUZUKI, H.; MAKAMURA, Y.; MURAKAMI, M.

Date: 860402

4081106

Adjustment of transformation point of shape memory alloys by introducing interstitial elements of, e.g., nickel-titanium alloys.

Patent Assignee: DAIDO TOKUSHUKO KK

Date: 860212

3. Piezo-Electric Materials

49560448

Translation generating device with piezo-electric actuator—has piezo-electric element stacks and support to transmit translation of actuator.

NoAbstract Dwg 1/2

Patent Assignee: (TOKE) TOSHIBA KK

Date: 870121

4687676

Piezo-electric pressure sensor—has series connected frequency discriminator, LF filter and rectifying diode.

Patent Assignee: (KHAV) KHARK AVIATION INST

Inventors: KOLESNIK, E. K.; SKULSKII, K. V.

Date: 850327

4642767

Piezo-electric ceramic electrode divider for AF sensor—divides piezo-electric ceramics by applying voltage between electrodes.

NoAbstract Dwg 1/5

Patent Assignee: (OMRO) OMRON TATEISI ELTRN KK

Date: 860307

4642765

Piezo-electric bimodal actuator—bonds two piezo-electric plates and intermediate metal plate.

NoAbstract Dwg 1/2

Patent Assignee: (TOHM) TOHOKU METAL IND LTD

Date: 860307

4634551

Piezo-electric sensor for motional feedback signal—has piezo-electric element composed of thin macromolecular piezo-electric film stuck on plate.

NoAbstract Dwg 1/16

Patent Assignee: (SONY) SONY CORP

Date: 860227

4573304

Testing of piezo-electric sensors—measuring capacitance during disconnection and connection of voltage source with serviceability is judged from change.

Patent Assignee: (VLAD/) VLADIMIROV, V. YA.

Inventors: VLADIMIROV, V. Y. A.; GOLDMAN, M. I.

Date: 840928

4530846

Piezo-electric pressure sensor has inertial body connected to vessel walls by elastic suspension preventing relative movement of electrodes during vibration.

Patent Assignee: (LEME=) LENGD MECH INST

Inventors: EROFEEV, N. K.; NOVIKOV, L. N.; SMIRNOV, A. P.

Date: 831207

4458368

Piezo-electric tangential stress sensor has lever element to transmit tangential component to piezo-elements whose deformation induces signals on electrodes.

Patent Assignee: (ASUR=) AS USSR URALS MECH

Inventors: BULAVIN, V. V.; PISTSOV, N. V.

Date: SU 850104

4340392

Piezo-electric actuator for stepping device has two actuators joined by electric distortion effect element fixed at opposite ends to faces of support.

Patent Assignee: (NIDE) NEC CORP

Date: 860507

4237835

Piezo-electric actuator comprises two ply piezo-electric element, increasing amount of translation.

NoAbstract Dwg 5/8

Patent Assignee: (MATU) MATSUSHITA ELEC IND KK

Date: 850122

4169756

Actuator with laminated piezo-electric ceramic driving source has body reciprocating in direction perpendicular to length of ceramic.

NoAbstract Dwg 1/5

Patent Assignee: (ASAH-) ASAHI OKUMA SANGYO

Date: 841026

4061853

Bar-type actuator with multiple drive piezo-electric elements applies voltage to each cut portion at one end of bar comprising piezo-electric elements.

NoAbstract Dwg 4/5

Patent Assignee: (HONF) YAMATAKE HONEYWELL KK

Date: 840703

4044421

Rate-acceleration sensor, e.g., for aircraft inertial reference has transducers generating signals in response to bending forces and a demodulation circuit reducing cross-axis coupling.

Patent Assignee: (ROCW) ROCKWELL INT CORP

Inventor: RIDER, B. F.

Date: 840113

4009608

Differential piezo-electric force sensor has spacer with arms of determined lengths of piezo-resonator and adjustable spring that applies static pulling forces to spacer.

Patent Assignee: (KHAV) KHARK AVIATION INST

Inventors: BARZHIN, V. Y. A.; KOLPAKOV, F. F.; MILKEVICH, E. A.

Date: 831102

3816796

Actuator has distortion of piezo-electric effect ceramic extracted as relatively large displacement.

Patent Assignee: (ASAH-) ASHIOHKUMA SANGYO

Date: 830930

3701143

Piezo-electric actuator lowers impressed voltage while maintaining expansion and contraction amount of material without using specified cooler.

Patent Assignee: (NIJI) NIPPON JIDOSHA BUHIN

Date: 830601

3251716

Highly sensitive piezo-electric accelerometer or sound sensor has weight resting centrally on piezo-electric device and supported by limiting webs.

Patent Assignee: (INKN-) indtech knufelmann

Inventor: KNUFELMANN, M.

3135858

Piezo-electric accelerometer has piezo-element with protrusions along perimeter engaging inertial mass and body elements to reduce lateral deformation body elements to reduce lateral deformation.

Patent Assignee: (SMIR/) SMIRNOV, V. V.

Inventor: SMIRNOV, V.

Date: SU 800505

2536434

Force feedback piezo-electric vibratory rate-of-turn sensor uses phase and magnitude of current supplied to transducer to determine rotation sense and rate of turn.

Patent Assignee: (SPER) SPERRY CORP

Inventor: JACOBSON, P. E.

Date: US 791001

4. Fiber Optics

5101943

Fibre optic sensor—having optical fibre formed into structural support coil that supports sensor material.

Patent Assignee: US SEC OF NAVY

Inventor: YUREK, A. M.

Date: 880415

5070024

Fibre optic sensor for environmental effects on counterpropagation—has detector connected beam-splitters to sense recombined beam and detects non-reciprocal phase shift to counterpropagating beam.

Patent Assignee: MCDONNELL DOUGLAS CORP

Inventors: UDD, E.; MICHAL, R. J.; WATANABE, S. F.; THERIAULT, P.; CAHILL, R. F.

Date: 861009

5035087

Microbend fibre optic sensor—has face to face microbending elements with ridges running transversely to coil that is sandwiched between.

Patent Assignee: UNIV OF MANCHESTER

Inventors: KVASNIK, F.; CAMPBELL, A. M.

Date: 870430

5026278

Fibre-optic sensor for measuring micro-displacements—uses photodetectors and optical grating to measure fringes on mirrors in pressure sensor.

Patent Assignee: PHOTNETICS

Inventors: ARDITTY, H.; VELLUET, M. T.

Date: 870316

5003802

Optical multiplexing system for fibre optic sensor—has common light source that feeds all sensors whose returns have individual filters duplicated at detector.

Patent Assignee: KOLLMORGEN CORP

Inventors: PARK, E. D.; GAT, E.

Date: 870413

4980091

Differential signal delay determination for fibre optic circuit determining difference between two frequencies on two signal paths to obtain representation of difference in signal propagation time.

Patent Assignee: LELAND STANFORD JR UNIV

Inventors: TUR, M.; KIM, B. Y.; BROOKS, J. L.; SHAW, H. J.

Date: 860623

4929868

Attaching an optical fibre to a metal surface—by metallising the fibre end and welding it to the metal surface using a current pulse.

Patent Assignee: BATTELLE MEMORIAL INST

Inventors: HOLMAN, R. L.; SKINNER, D. P.; JOHNSON, L. M.

Date: 880122

4929848

Fibre optic sensor responsive to force or pressure—utilises phase shift induced by deformation of optical fibre embedded in block of creep-resistant elastomer.

Patent Assignee: PFISTER GMBH; HAFNER, H. W.

Inventor: HAFNER, H. W.

Date: 870121

4843785

Fibre optic sensor for measurement of minimum elongation—provides evaluation of phase difference between components of propagation in birefringent fibre with directionally-stressed sheath.

Patent Assignee: FELTEN 7 GUIL ENERG

Inventors: FEDERMANN, H.; LEVACHER, F. K.; NOACK, G.; KRAUS, A.

Date: 870829

4814689

Crossed fibre optic tactile sensor—has column of detecting fibres to characterise object by measuring spatial distribution of contact forces involved in gripping operation.

Patent Assignee: ROCKWELL INT CORP

Inventors: THIELE, A. W.; SCHOENWAL, J. S.; GJELLUM, D. E.

Date: 860407

4781076

Crack detection system for aircraft bodies—has fibre-optic sensors distributed through body and cyclically monitored by cockpit computer.

Patent Assignee: (MESR) MESSERSCHMITT-BOLKOW-BLO

Inventors: HOFER, B.; RONNER, C.; MALEK, S.

Date: 860829

471884

Fibre optic sensor for detecting ground-fault on power cable—uses optical fibres with thin and thick clad fibre sections alternately formed.

NoAbstract Dwg 4/7

Patent Assignee: HITACHI CABLE KK

Date: 860520

447944

Fibre optic sensor for temperature pressure and physical parameters has probe tip with luminescing coating backed by clear elastomeric cushion deformable under contact.

Patent Assignee: LUXTRON CORP

Inventors: SUN, M.; WICKERSHEI, K. A.; HEINEMANN, S. G.

Date: 861022

4424566

Fibre-optic sensor for measuring temperature or strain variation uses two fibres with equal phase changes for variation of non-measured variables.

Patent Assignee: (DAII-) DAIICHI DENSHI KOGY

Inventors: HIROSE, T.; MATSUMURA, Y.

Date: 850903

4410062

Piezo-electric phase modulator for fibre optic sensor has fibre bonded to relatively stable material as part of optic sensor loop to produce mechanical vibration under excitation.

Patent Assignee: (UNAC) UNITED TECHNOLOGIES CORP

Inventors: FOURNIER, J. T.; SWARTS, R. E.

Date: 850822

4410031

Fibre optic sensor for physical parameter measurement has temperature sensitive material to control degree of light transmission between ends of two optical fibres within sensing head.

Patent Assignee: (TACA-) TACAN AEROSPACE CORP

Inventors: SALOUR, M. M.; SCHONER, G.; BECHTEL, J. H.; KULL, M.

Date: 850819

4380319

Microwatt battery powered fibre optic sensor, e.g., for temperature and pressure utilises LED driven by optical transducer and biased by battery at level that exceeds shot noise level of LED by dynamic range of sensor.

Patent Assignee: (SPER) SPERRY CORP

Inventors: McMAHON, D. H.; SOREF, R. A.

Date: 840713

4317233

Wavelength switch passive interferometer for fibre optic sensor has optical fibre and sensing signal producing phase difference in response to physical change.

Patent Assignee: (LITO) LITTON SYSTEMS INC

Inventor: (LITO) LITTON SYSTEMS, INC.

Inventor: BUSH, I. J.

Date: 860319

4242364

Fibre optic ultrasound detector has fibre with cladding that exhibits relatively higher refractive change with pressure than core.

Patent Assignee: (CANA) NAT RES COUNCIL CAN

Inventor: CIELO P. G.

Date: 830120

4216673

Differential fibre optic sensor for copier has light source attached to substrate to which transmission and reception optical waveguides are secured.

Patent Assignee: (XERO) XEROX CORP

Inventor: JEDLICKA, J. E.

Date: 860205

4216671

Distributed parameter sensing system has optical source with short coherence length for continuously monitoring each sensor of the system.

Patent Assignee: (STRD) LELAND STANFORD JR UNIV

Inventors: YOUNGQUIST, R. C.; BROOKS, J. L.; FESLER, K. A.; CUTLER, C. C.; SHAW, H. J.

Date: 860205

4216670

Distributed sensor system with optical source uses coherence multiplexing of fibre-optic interferometric sensors.

Patent Assignee: (STRD) LELAND STANFORD JR UNIV

Inventors: BROOKS, J. L.; TUR, M.; YOUNGQUIST, R. C.; KIM, B. Y.; WENTWORTH, R. H.; SHAW, H. J.; BLOTEKJAER, K.

Date: 860205

4158902

Fibre optic sensor lead fibre noise cancellation has beam splitter to provide two interference patterns sensed by two photodetectors.

Patent Assignee: (DRAS) STARK DRAPER C LAB INC

Inventor: CHAMUEL, J. R.

Date: 830519

4019856

Fibre optic sensor for temperature, pressure, etc. has light source modulated by superimposing ac from source on laser drive also applied to comparator.

Patent Assignee: (STTE) STAND TEL & CABLES PLC

Inventor: JONES, R. E.; PRATT, R. H.

Date: 840720

3937668

Fibre optic sensor using two parallel optical fibres touching at coupling point where sheat may be removed to expose cause of cross transmission.

Patent Assignee: (FRAU) FRAUNHOFER-GES FORD ANGE

Inventors: RAMAKRISH, S.; KERSTEN, R.

Date: 840424

3832113

Fibre optic sensor for remote or hostile environment simultaneously detects different parameters including temperature, pressure and voltage in single sensing tip.

Patent Assignee: (AETN-) AETNA TELECOMMUNI

Inventor: NELSON, A. R.

Date: 820729

3827610

Polymeric optical environmental change detector in which polymer changes due to environmental changes are detected optically.

Patent Assignee: (MONS) MONSANTO CO

Inventor: FISHER, W. K.; WALKER, G. E.

Date: 841204

3603112

Protective fibre optic sensor for measuring force and pressure has monomode fibre deformed only slightly and has change in polarisation evaluated.

Patent Assignee: (LICN) LICENTIA PATENT GMBH

Inventor: WINTERHOFF, H.

Date: 831221

3553778

Fibre optic sensor detecting changes in physical quality, e.g., press includes parallel interfaces separated by changeable gap causing interference patterns in incident and photoluminescent optical energies.

Patent Assignee: (ALLM) ASEA AB

Inventor: BROGARDH, T.; HOK, B.; OVREN, C.

Date: 831117; US 831121

3361860

Fibre optic sensing device, e.g., for temperature receives optical radiation back scattered by length of liquid-core optical fibre measuring transmission intensity; OPTICAL TIME DOMAIN REFLECTOMETER.

Patent Assignee: (NATR) NATIONAL RES DEV CORP

Inventors: HARTOG, A. H.; PAYNE, D. N.

Date: 820518

3349819

Fibre optic sensor measuring dynamic movement of surface has photoluminescent material in end of optical fibre to measure relative movement of body.

Patent Assignee: (ALIM) ASEA AB

Inventors: BROGARDH, T.; HOK, B.; OVREN, C.

Date: 830526

3343913

Fibre optic sensor measuring physical qualities had distributed luminescent centres responding to variable being measured.

Patent Assignee: (ALLUM) ASEA AB

Inventors: BROGARDH, T.; HOK, B.; OVREN, C.

Date: 830526

3300500

Fibre optic sensor for surface displacement detection has two fibres of different length between directional couplers, connected to detector and providing movement over the surface; LIGHT MODULATE PHASE MAGNETIC DISC STORAGE.

Patent Assignee: (STRI) LELAND STANFORD JR UNIV

Inventors: KINO, G. S.; BOWERS, J. E.

Date: 820414

APPENDIX B: BIBLIOGRAPHY

1. Electro-Rheological Fluids

SCOTT, D. and YAMAGUCHI, J., "ER Fluid Devices Near Commercial Stage," *Automotive Engineering*, November 1985, Vol. 95, No. 11, pp. 75–79

BULLOUGH, W. A., PEEL, D. J. and FIROOZIAN, R., "Electrical Characteristics and Other Considerations for Electro-Rheological Fluids," *Proceedings of the First International Symposium on ER Fluids, 18th Annual Meeting of the Fine Particle Society, Boston, 1987*

PEEL, D. J., FIROOZIAN, R. and BULLOUGH, W. A., "Deviation of the Electrical Characteristics of an Electro-Rheological Fluid from Biased Sine Wave Excitation Tests," *ibid.*

ARGUELLES, J., MARTIN, H. R. and PICK, R., "A Theoretical Model for Steady Electroviscous Flow between Parallel Plates," I Mech. E. *Journal of Mechanical Engineering Science*, Vol. 16, No. 4, 1974, pp. 232–239

SCHULMAN, Z. P., GORODKIN, R. G., KOROBKO, E. V. and GLEB, V. K., "The Electro-Rheological Effect and Its Possible Uses," *Journal of Non-Newtonian Fluid Mechanics*, Vol. 8, 1981, pp. 29–41

DUCLOS, T. G., ACKER, D. N. and CARLSON, J. D., "Fluids that Thicken Electrically," *Machine Design*, January 21, 1988, pp. 42–46

HINCH, E. J. and SHERWOOD, J. D., "The Primary Electroviscous Effect in a Suspension of Spheres with Thin Double Layers," *Journal of Fluid Mechanics*, 1983, Vol. 132, pp. 337–347

HONDA, T., KUROSAWA, K. and SASADA, T., "A. C. Characteristics of the Electroviscous Effect," *Japanese Journal of Applied Physics*, Vol. 19, No. 8, 1980, pp. 1463–1466

GORODKIN, R. G., KOVOBKO, Y. V., BLOKH, G. M., BLEB, V. K., SIDOROVA, G. I. and RAGOTHER, M., "Applications of the Electro-Rheological Effect in Engineering Practice," *Fluid Mechanics—Soviet Research*, Vol. 8, No. 4, 1979, pp. 48–61

UEJIMA, H., "Dielectric Mechanism and Rheological Properties of Electro-Fluids," *Japanese Journal of Applied Physics*, Vol. 11, No. 3, 1972, pp. 319–325

MULLINS, J. P., "Fluid Powerful!", *Automotive Industries*, September 1988, pp. 68–69

VINGRADOV, G. V., SHULMAN, Z. P., YANOVSKI, Y. G., BARANCHEEVA, V. V., KOROBKO, E. V. and BUKOVICH, I. V., *Inzhenerno-Fizicheskii Zhurnal*, Vol. 50, No. 4, 1986, pp. 605–609

SHULMAN, Z. P., KHUSID, B. M., KOROBKO, E. V. and KHIZHINSKY, E. P., "Damping of Mechanical Systems Oscillation by a Non-Newtonian Fluid with Electric-Field Dependent Parameters," *Journal of Non-Newtonian Fluid Mechanics*, Vol. 25, 1987, pp. 329–346

ANASKIN, I. F., GLEB, V. K., KOROBKO, E. V., KHIZHINSKII, B. P. and KHUSID, B. M., "Effect of External Electric Field on Amplitude-Frequency Characteristics of Electro-Rheological Damper," *Inzhenerno-Fizicheskii Zhurnal*, Vol. 46, No. 2, 1984, pp. 309–315

BULLOUGH, W. A., "A Simple Drag Pump," *Proc. 4th Int. Conf. Fluid Machinery, Budapest*, pp. 211–226, 1972: Academy of Sciences, Budapest

BULLOUGH, W. A. and STRINGER, J. D., "The Utilisation of the Electroviscous Effect in a Fluid Power System," *Proc. 3rd Int. Fluid Power Symposium, Turin*, pp. 371–451, 1973: British Hydromechanics Research Association

BULLOUGH, W. A. and FOXON, M. B., "The Application of an Electroviscous Damper to a Vehicle Suspension System," *Proc. 3rd Int. Conf. Vehicle System Dynamics, Blacksburg, Virginia*, pp. 144–160, 1974: Swets and Zeitlinger, Amsterdam

BULLOUGH, W. A., "The Design of a Glandless Oil Pump," *Proc. 5th Int. Conf. Fluid Machinery, Budapest*, pp. 135–147, 1975: Academy of Sciences, Budapest

BULLOUGH, W. A. and J. D. STRINGER, "Direct Flow Control by Electric Fields," Hydraulics and Air Power, *Jnl. Applied Hydraulics and Pneumatics Tech.*, Vol. 5, No. 4, pp. 11–17, 1977: Applied Technology Publications Limited

BULLOUGH, W. A. and FOXON, M. B., "A Proportionate Coulomb and Viscously Damped Isolation System," *Jnl. Sound and Vibration*, Vol. 56, No. 1, pp. 1–10, 1978: Academic Press

BULLOUGH, W. A., "Testing HGMS Mesh," *Jnl. Filtration and Separation*, Vol. 15, No. 4, pp. 303–305, July, 1978

BULLOUGH, W. A., "HGMS in the Region of High Aerodynamic Loading," *Jnl. Applied Phys.,* Vol. 50, pp. 3057–3062, May, 1979

BULLOUGH, W. A., "Glandless Pump Constructions for Low Duty Applications," *Proc. 6th Int. Conf. Fluid Machinery, Budapest*, pp. 146–160, 1979: Academy of Sciences, Budapest

BULLOUGH, W. A., "The Flow Characterisation of a High Gradient Magnetic Separator," *Jnl. Inst. Phys.*, Vol. 12D, pp. 335–343, 1979

BULLOUGH, W. A., "The Influence of Gasket Tolerance on a Pump Loaded with Flanged Pipe Lengths," *Proc. 7th Int. Conf. Fluid Machinery, Budapest*, pp. 96–105, 1983: Academy of Sciences, Budapest

BULLOUGH, W. A. and D. J. PEEL, "Electro-Rheological Oil Hydraulics," *Journal of the Japan Society for Hydraulics and Pneumatics*, Vol. 17, pp. 30–36, 1986

BULLOUGH, W. A., "Electron Hydraulics—A Series of Lectures," Published by Beijing Inst. of Tech. 61 pp. (in Chinese)

BULLOUGH, W. A., PEEL, D. J. and FIROOZIAN, R. "Electrical Characterisation and Other Considerations for Electro-Rheological Fluids," 1st Int. Symp. on Electro-Rheological Fluids, Boston, Mass. The Fine Particle Society. Accepted for publication by the American Physical Society

BULLOUGH, W. A. "Poppet Valve Studies," 8th Int. Conf. on Fluid Machinery, Budapest, pp. 112–119, 1987: Academy of Sciences, Budapest

PEEL, D., FIROOZIAN, R. and BULLOUGH, W. A., "The Derivation of the Electrical Characteristics of an Electro-Rheological Fluid from Biased Sine Wave Excitation Tests," *Proc. 8th Intl. Fluid Power Symposium, NEC Birmingham*, 17 pp., 1988: British Hydrodynamics Research Association

BULLOUGH, W. A., "Integrated Electron-Hydraulics," *Proc. of Mech. 88, World Congress of the Australian Bicentennial, Exposition, Brisbane*, 6 pp., 1988: I. Mech. E., ASME Japan SME Inst. of Eng. Australia

BULLOUGH, W. A., "Present Trends and Future Research Directions in Electro-Rheological Fluids," International Heat and Mass Transfer forum, Minsk, 24–27 May, 1988. To be published in the proceedings. BSSR Academy of Science

BULLOUGH, W. A. and FIROOZIAN, R. "Integrated Piezo Technology at the Electron-Hydraulic Interface," International Heat and Mass Transfer Forum, Minsk, 24–27 May, 1988. To be published in the proceedings. BSSR Academy of Science

BULLOUGH, W. A. and PEEL, D. J., Electro-Rheological Oil Hydraulics," *Proceedings of Japan Society for Hydraulics and Pneumatics*, Vol. 61, No. 17, pp. 520–526, November 1986

FIROOZIAN, R. "A Comparison between Hydraulic and Electrical d.c. Servo Motor Feed-Drive Systems," M.Sc. dissertation, The University of Aston in Birmingham, October 1979

FIROOZIAN, R., "The Comparison of the Performance of Servo Motor Feed-Drive Systems," Ph.D. thesis, The University of Aston in Birmingham, 1983

FIROOZIAN, R., BRIMSON, J. R. and FOSTER, K. "The Comparison of Performance of Servo Feed-Drive Systems," I. Mech. E. Conf. on Electric versus Hydraulics Drives, Mechanical Engineering Publications Ltd., London, 27 October 1983

FIROOZIAN, R. and FOSTER, K., "The Choice of a Servo Motor for a Specific Application," I. Mech. E. Conf. on Electrics versus Hydraulics versus Pneumatics, Mechanical Engineering Publications Ltd., London, 22 October 1985, pp. 67–76

FIROOZIAN, R. and STANWAY, R., "Influence of Suspension Parameters on Vibration Control in Machinery," ASME Conf. on Mechanical Vibration and Noise, Cincinnati, 1985, Paper 85-DET-125.

STANWAY, R., FIROOZIAN, R. and MOTTERSHEAD, J. E., "Identification of the Linearized Dynamics of a Squeeze-Film Vibration Isolator," *Proc. Instn. Mech. Engrs.*, Vol. 201, No. C2, 1987

STANWAY, R., SPROSTON, J. L. and FIROOZIAN, R. "Identification of the Damping Law of an Electro-Rheological Fluid in Vibration," ASME Winter Annual Meeting, Anaheim, California, December 1986, paper 86-WA/DSC-6

STANWAY, R., MOTTERSHEAD, J. E. and FIROOZIAN, R., "Non-linear Identification of a Squeeze-Film Damper," 4th Workshop on Rotordynamics Instability Problems in High Performance Turbo-Machinery, Texas A and M University, June 1986 (for publication in Journal of Tribology)

FIROOZIAN, R. and STANWAY, R., "Modelling and Control of Turbo-Machinery Vibration," ASME Design Engineering Division Conference on Vibration and Noise, 27–30 September, 1987, Boston, USA

STANWAY, R., FIROOZIAN, R. and MOTTERSHEAD, J. E., Implication of Modal Reduction in the Active Stabilisation of Turbo-Machinery," Euromech 213, Active Methods for Controlling Sound and Vibrations, Marseille, France, September 8–11, 1986

STANWAY, R., MOTTERSHEAD, J. E. and FIROOZIAN, R., "System Modelling and Its Role in CAE of Mechanical Structures," CADCAM CONFERENCE, 8–10 April, 1986, NEC Birmingham, pp. 199–205

MOTTERSHEAD, J. E., FIROOZIAN, R. and STANWAY, R., "Identification of Damping Parameters of Sqeeze-Film Damper. Time Domain or Frequency Domain?", Leeds/Lyon Conference on Tribology, 8–11 September 1986, Leeds

BULLOUGH, W. A., PEEL, D. J. and FIROOZIAN, R., "Electric Characteristics and Other Considerations for Electro-Rheological Fluids," 18th Annual Meeting of the Fine Particle Science, August 1987, Boston, USA. Submitted for publication in the *International Journal of the Fine Particle Society*

PEEL, D. J., FIROOZIAN, R. and BULLOUGH, W. A., "Derivation of the Electrical Characteristics of an Electro-Rheological Fluid from Biased Sine Wave Excitation Tests," accepted for presentation at the 8th International Symposium on Fluid Power, 18–21 April 1988, Birmingham

FIROOZIAN, R., "Implication of Model Reduction in the Active Control of Turbomachinery Vibrations." Active Noise and Vibration Control, special issue supplement to Vol. 6, 1987, *Journal De Mechanique Theorique et Appliquee (Journal of Theoretical and Applied Mechanics)*, pp. 183–202

FIROOZIAN, R., "Condition Monitoring and Diagnostic Analysis of Rotating Machinery Using Parameter Identification." Accepted for presentation at COMADEM 88—First UK Seminar, Birmingham Polytechnic

FIROOZIAN, R. and STANWAY, R., "Active Vibration Control of Turbomachinery: A Numerical Investigation of Modal Controllers." Accepted for publication in *ASME Journal of Vibration, Acoustic, Stress and Reliability in Design* (to appear)

FIROOZIAN, R. and STANWAY, R., "Design and Application of a Finite-Element Package for Modelling Turbomachinery Vibrations." Accepted for publication in *Journal of Sound and Vibration*

WINSLOW, W. M., "Induced Vibration of Suspensions," *J. Appl. Phys.* 1949

ARGUELLES, J., MARTIN, H. and PICK, R., "Some Experiments with Electrosensitive Fluids," Third International Fluid Power Symposium 1973 (Turin)

ANDRADE, E. N. and DODD, C., "The Effect of an Electric Field on the Viscosity of Liquids," *Proc. R. Soc.* 1949 1987, 1971 201A, 1954 225A

ELTON, G. A. H. and HIRSCHLER, F. G., "Electroviscosity," *Proc. R. Soc.* 1948 194, 1949 197, 1949 198

SAITOH, S., SAKAI, T. and KUSAMA, H. "Fluidic Electro-Fluid Converter by Use of Electro-Viscous Fluid," Second International JSME Symposium September 1972 (Tokyo)

VELKOFF, H. R., "Evaluating the Interactions of Electrostatic Fields with Fluid Flows," ASME Design Engineering Conference April 1971

STRANDRAT, H. T., *Proc. Hydr. and Pneumet.*, 1966, pp. 139–143

SHULMAN, Z. P., MATSEPURO, A. D. and SMOLSKY, B. M., *Vesti An BSSR, Ser. Fiz, Energ. Nauk*, Vol. 1 (1974), p. 60

SHULMAN, Z. P., MATSEPURO, A. D. and KUZMIN, V. A., *Vesti AN BSSR, Ser. Fiz. Energ. Nauk*, Vol. 2 (1974), p. 130

SHULMAN, Z. P. and KOROBKO, E. V., *Inst. J. Heat Mass Transfer*, Vol. 21 (1978) p. 543

LUIKOV, A. V., Electro-Rheological Effect. *Nauka i Tekhn. Minsk*, 1972

BATCHELOR, G. K., 1970 *J. Fluid Mech.*, Vol. 41, p. 545

BOOTH, F., 1950 *Proc. R. Soc. Lond.*, Vol. A203, p. 533

DUKHIN, S. S. and DERJAGUIN, B. V., 1974 in *Surface and Colloid Science, Vol. 7* (ed., E. Matijevic), chap. 3, Wiley

LEVER, D. A., 1979, *J. Fluid Mech.*, Vol. 92, p. 421

RUSSEL, W. B., 1978, *J. Fluid Mech.*, Vol. 85, p. 209

SHERWOOD, J. D., 1980, *J. Fluid Mech.*, Vol. 101, p. 609

WATTERSON, I. G. and WHITE, L. R., 1981, *J. Chem. Soc. Faraday Trans. II*, Vol. 77, p. 1115

SOKOLOV, P. and SOSINSKI, S., *Acta Phisicochim. (URSS)*, Vol. 5 (1936), p. 691

DA, E. N., ANDRADE, C. and DODD, C.: *Proc. R. Soc. London*, Vol. A187 (1946), p. 294

HONDA, T. and ALLEN, P., *Proc. 5th Int. Conf. Conduction and Breakdown in Dielectric Liquids* (Delft Univ. Press, 1975), p. 131

HONDA, T. and ALLEN, P., *Proc. 6th Int. Conf. Conduction and Breakdown in Dielectric Liquids* (Edition Frontieres, 1978), p. 131

HONDA, T. and SASADA, T., *Jpn. J. Appl. Phys*, Vol. 16 (1977), pp. 1775, 1785

HONDA, T., KUROSAWA, K. and SASADA, T., *Jpn. J. Appl. Phys.*, Vol. 18 (1979), pp. 2095

HONDA, T. and SASADA, T., *Jpn. J. Appl. Phys.*, Vol. 18 (1979), p. 675

HONDA, T. and SASADA, T., *Jpn. J. Appl. Phys.*, Vol. 18 (1979), p. 1605

REINER, M. A., *Lectures in Theoretical Rheology*, North Holland, Amsterdam, 1960 [Rus. transl. as *Reologiya (Rheology)*, 1965].

REINER, M. A., "Mathematical Theory of Dilatancy," *Amer. J. Math.*, Vol. 67 (1945), p. 350

ANON., "Fluid Makes a Viscosity Pump when Jarred by an Electric Field," *Machine Design*, May 24, 1962

ANON. "Electro-Fluids for Power Transmission," *Electrical Manufacturing*, Vol. 63, No. 2, p. 11, 1959

"Dirty Oil: The Discovery and Application of the Winslow Effect," *Hydraulics and Pneumatics*, Vol. 19 (1962), pp. 144–147

STRAADRUD, H. T., "Electric Field Valves Inside Cylinder Control Vibrator," *Ibid.*, Vol. 19, pp. 139–143

PHILLIPS, R. W. and AUSLANDER, D. W., "The Electroplastic Flow Modulator," Source uncertain, May, 1971

PHILLIPS, R. W., "Engineering Applications of Fluids with a Variable Yield Stress," Dissertation, University of California, 1969

SMOLUCHOWSKI, M. V., *Kolloid-Z*, Vol. 18 (1916), p. 190

KRANSY-ERGEN, W., *Kolloid-Z*, Vol. 74 (1936), p. 172

BOOTH, F., *Proc. Roy. Soc.*, Vol. A203 (1950), p. 533

HARMSEN, G. J., *J. Colloid. Sci.*, Vol. 8 (1953), pp. 64, 72

BJORNSTÀHL, Y. and SNELLMAN, K. O., *Kolloid-Z*, Vol. 78 (1958), p. 258

KLASS, D. L. and MARTINEK, T. W., *J. Appl. Phys.*, Vol. 38 (1967), p. 67

KATO, I., ET AL., *Transaction of Institute of Electrical Engineerings of Japan at Tokyo Prefectue* (1966), p. 66 (in Japanese)

KATO, I., ET AL., *Transaction of Institute of Electrical Engineers of Japan at Tokyo Prefecture* (1968), p. 122 (in Japanese)

SCHWARZ, G., *J. Phys. Chem.*, Vol. 66 (1962), p. 2636

ANASKIN, I. F., ILEB, V. K., KHUSID, B. M., KOROBKO, E. V. and KHIZHIUSKY, B. P., *J. Eng. Phys. (Russian)*, Vol. 26 (1984), p. 309

ZHOKHOVSKY, M. K., "Theory and Design of Devices with an Unsealed Piston," *Izd. Standartov*, Moscow, 1980

SHULMAN, Z. P., MATSEPURO, A. D. and KHUSID, B. M., *Izv. AN BSSR, Ser. Fix. Energ. Nauk*, Vol. 3 (1977), p. 116

SHULMAN, Z. P., MATSEPURO, A. D. and KHUSID, B. M., *Izv. AN BSSR, Ser. Fiz, Engerg. Nauk*, Vol. 3 (1977), p. 122

LUIKOV, A. V. (ed), *Electro-Rheological Effect*, Nauka i Tekhnika, Minsk, 1972

SHULMAN, Z. P., MATSEPURO, A. D. and KHUSID, B. M., "Structurization of Electro-Rheological Suspensions in Electric Field, Part 1: Qualitative Analysis," *Izv. Akad. Nauk BSSR Ser. Fix.-Energ. Nauk*, No. 3, pp. 116–121 (1977)

SHULMAN, Z. P., MATSEPURO, A. D. and KHUSID, B. M., "Structurization of Electro-Rheological Suspensions in Electric Field, Part 2: Quantitative Estimates," *Izv. Akad. Nauk BSSR, Ser. Fix-Energ. Nauk*, No. 3, pp. 122–127 (1977)

SHULMAN, Z. P., *Electro-Rheological Effect and Its Possible Applications (in Russian)*, Inst. Teplo-Masso Obmena, Minsk (1975)

JONES, B. E., *Electrotechnology (GB)*, (Oct. 1984), Vol. 12, No. 4, pp. 148–155

THOMPSON, B. S. and M. V. GANDHI, "A Commentary of Flexible Fixturing," *ASME Applied Mechanics Reviews*, Vol. 39, No. 9, 1986, pp. 1365–1369

GANDHI, M. V., THOMPSON, B. S., CHOI, S. B. and SHAKIR, S. "Electro-Rheological-Fluid-Based Articulating Robotic Systems," ASME Paper No. 87-DAC-55, published in *Advances Design Automation—1987: Volume Two, Robotics, Mechanisms, and Machine Systems*, (S. S. Rao, ed.) DE-Vol. 10-2, pp. 1–10, and *ASME Journal of Mechanisms, Transmissions and Automation in Design* (in press)

CHOI, S. B., THOMPSON, B. S. and GANDHI, M. V., "An Experimental Investigation on the Active-Damping Characteristics of a Class of Ultra-Advanced Intelligent Composite Materials Featuring Electro-Rheological Fluids," *Proceedings of the Damping '89 Conference, West Palm Beach, February 1989*, organized by the Flight Dynamics Laboratory of the Air Force Wright Aeronautical Laboratories, Wright Patterson Air Base, Ohio

CHOI, S. B., GANDHI, M. V. and THOMPSON, B. S., "An Active Vibration Tuning Methodology for Smart Flexible Structures Incorporating Electro-Rheological Fluids: A Proof-of-Concept Investigation," 1989 Automatic Control Conference, Pittsburgh, June 1989

CHOI, S. B., GANDHI, M. V. and THOMPSON, B. S., "An Experimental Investigation of the Elastodynamic Response of a Slider-Crank Mechanism Featuring a Smart Connecting Rod Incorporating Embedded Electro-Rheological Fluid Domains," to be presented at the 1st National Mechanisms Conference, Cincinnati, Ohio, Nov. 1989

THOMPSON, B. S. and GANDHI, M. V., "A Variational Formulation for Smart Links of Robotic and Mechanism Systems Featuring Embedded Electro-Rheological Fluids," to be presented at the 1st National Mechanisms Conference, Cincinnati, Ohio, Nov. 1989

GANDHI, M. V. and THOMPSON, B. S., "An Innovative Class of Revolutionary Articulating Mechanism and Robotic Systems Exploiting Smart Materials and Structures: Fiber Optics, Shape Memory Metals, Piezoelectric Materials and Electro-Rheological Fluids," to be presented at the 1st National Mechanisms Conference, Cincinnati, Ohio, Nov. 1989

CHOI, S. B., THOMPSON, B. S. and GANDHI, M. V., "Smart Structures Incorporating Electro-Rheological Fluids for Vibration-Control and Active-Damping Applications: An Experimental Investigation," to be presented at the 12th Biennial ASME Conference on Mechanical Vibration and Noise, Montreal, Sept. 1989, and *ASME Journal of Vibration, Acoustics, Stress and Reliability in Design* (under review).

GANDHI, M. V., THOMPSON, B. S. and CHOI, S. B. "A New Generation of Innovative Ultra-Advanced Intelligent Composite Materials Featuring Electro-Rheological Fluids: An Experimental Investigation," *Journal of Composite Materials* (in press).

CHOI, S. B., GANDHI, M. V. and THOMPSON, B. S. "An Active Vibration Tuning Methodology for Smart Flexible Structures Incorporating Electro-Rheological Fluids: A Proof-of-Concept Investigation," *Journal of Guidance, Control and Dynamics* (under review).

"A Note on an Experimental Investigation of a Slider-Crank Mechanism Featuring an Smart Dynamically-Tunable Connecting-Rod Incorporating and Electro-Rheological Fluid; *Journal of Sound and Vibration* (under review).

"Active Vibration Tuning of a Simply-Supported Beam Featuring Electro-Rheological Fluids," *ASME Journal of Dynamic Systems, Measurement, and Controls.*

GANDHI, M. V., THOMPSON, B. S. and CHOI, S. B., "Smart Ultra-Advanced Composite Materials Incorporating Electro-Rheological Fluids for Military, Aerospace and Automated Manufacturing Applications," SME/EDS Advanced Composites Conference and Exposition, published in *Advanced Composites III: Expanding the Technology*, pp. 23– 30, Detroit, MI, September 1987.

CHOI, S. B., THOMPSON, B. S. and GANDHI, M. V., "Electro-Rheological Fluids Technology Stimulates a New Generation of Robotic and Machine Systems," *Proceedings of the Tenth OSU Applied Mechanisms Conference, New Orleans, December 1987*, Vol. I, pp. 3B.2-1 to 3B.2-8.

GANDHI, M. V., THOMPSON, B. S. and CHOI, S. B., "Ultra-Advanced Composite Materials Incorporating Electro-Rheological Fluids," *Proceedings of the 4th Japan-U.S. Conference on Composite Materials, Washington, D.C., June 1988*, pp. 875–884.

GANDHI, M. V. and THOMPSON, B. S., "A New Generation of Revolutionary Intelligent Composite Materials Featuring Electro-Rheological Fluids," *Proceedings of the ARO Workshop on Smart Materials, Structures and Mathematical Issues, Blacksburg, VA, September, 1988.*

2. Shape Memory Alloys

"Shape Memory Metals."
KUBEL, JR., E. J.
Mater. Eng. (Cleveland), Vol. 99, No. 6, pp. 27–30, June 1984

The Design and Testing of a Memory Metal Actuated Boom Release Mechanism.
POWLEY, D. G. and BROOK, G. B.
12th Aerospace Mechanism Symposium, Moffett Field, Calif., 27–28 Apr. 1978
Publ.: NASA Conference Pub. 2080. National Aeronautics and Space Administration, Washington, D.C. 20546, 1979, pp. 119–129

"Out of the Metallurgists' Box of Tricks: Alloys with Shape Memory."
ASCHMONEIT, E.-K.
VDI Nachr., Vol. 40, No. 10, Magazin, pp. 36–37, 7 Mar. 1986

"Mechanical Behavior of the Cu—Zn—Al Shape Memory Alloys."
LI, K. and CAO, M. S.
Southwest Aluminium Fabrication Plant, Cantral-South University of Technology
Proceedings of the International Conference on Martensitic Transformations. ICOMAT-86, Nara, Japan, 26–30 Aug. 1986
Publ.: The Japan Institute of Metals, Aoba Aramaki, Sendai 980, Japan, 1987, pp. 914–919

"Alloys with Shape Memory—New Materials for Future Technology?"
HORBOGEN, E.
Metall. Vol. 41, No. 5, pp. 488–493, May 1987

"Effect of Heat Treatment on Martensite Transformation and Shape Memory in Brasses with Aluminum and Tin Additives."
DUTKIEWICZ, J. and MORGIEL, J.
International Conference Non-Ferrous Metals '87, Krakow, Poland, 1987 Metal. Odlew. (109), pp. 563–572 1987 ISSN: 0137-6536

"Structure and Shape Memory in Cu-Zn-Al Alloy."

BOJARSKI, Z., MORAWIEC, H., LELATKO, J. and AUGUSTYNIAK, M.

Pr. Inst. Met. Niezelaz. Vol. 13, Nos. 1–2, 111–120, 1984

"Phase Transitions in Shape Memory Alloys and Implications for Thermal Energy Conversion".

BANKS, R.

Publ.: Elsevier Applied Science Publishers Ltd., Crown House, Linton Rod., Barking, Essex IG11 8JU, UK, 1986

Phase Transformations, pp. 77–96

"Shape Memory Effect Alloys: Basic Principles."

PERKINS, J.

Publ.: Pergamon Press, Headington Hill Hall, Oxford OX3 OBW, UK, 1986

Encyclopedia of Materials Science and Engineering, Vol. 6 (Met. A., 8609-72-0335) 4365–4368

"Iron–Nickel–Titanium–Cobalt Alloy with Shape Memory Effect and Pseudo-Elasticity, and Method of Producing the Same."

TAMURA, I. and MAKI, T.

Kyoto University

Ausz. Eur. Patentanmeld. I 2, (2), p. 92, 8 Jan. 1986

"Rapidly Quenched Shape Memory Alloys."

EUKEN, S. and HORNBOGEN, E.

Rapidly Quenched Metals, Vol. II, Wurzburg, FRG, 3–7 Sept. 1984

Pub.: Elsevier Science Publishers B. V., P.O. Box 103, 1000 AC Amsterdam, The Netherlands, 1985

"Creep of Copper-Base Shape Memory Alloys."

VERGUTS, H., AERONOUDT, E. and DELAEY, L.

Strength of Metals and Alloys (ICSMA 7). Vol. 3, Montreal, Canada, 12–16 Aug. 1986

Publ.: Pergamon Press Ltd., Headington Hill Hall, Oxford OX3, OBW, UK, 1985

"Microstructure and Mechanical Properties of Rapidly Quenched Shape Memory Alloy."

EUKEN, S. and HORNBOGEN, E.

Strength of Metals and Alloys (ICSMA 7). Vol. 2, Montreal, Canada, 12–16 Aug. 1985

Publ.: Pergamon Press Ltd., Headington Hill Hall, Oxford OX3 OBW, UK, 1985

"Monitoring the Response of Shape Memory Alloys Along Thermomechanical Paths in Transformation Space."

VERGUTS, H. and AERONOUDT, E.

Strength of Metals and Alloys, Vol. 1, Montreal, Canada, 12–16 Aug. 1985

Publ.: Pergamon Press Ltd., Headington Hill Hall, Oxford OX3 OBW, UK, 1985

"Phase Transitions in the Intermetallic Compound TiNi with Charge-Density Wave Formation."

SHABALOVSKAYA, S. A.

Phys. Status Solidi (b) 132, (2), 327–344 Dec. 1985

"A Method for Designing Shape Memory Alloy Springs."

SAKAI, K.

Mitsubishi Steel

Mitsubishi Steel Manuf. Tech. Rev., Vol. 20, Nos. 1–2, pp. 7–12, 1986

"Composites Containing Shape Memory Fibers."

THUMANN, M., VELTEN, B. and HORNBOGEN, E.

Ruhr-Universitat Bochum

Proceedings of the International Conference on Martensitic Transformations. ICOMATO-86, Nara, Japan, 26–30 Aug. 1986

Publ.: The Japan Institute of Metals, Aoba Aramaki, Sendai 980, Japan 1987

"Behavior of Reversible SMA Coupler for Quick Replacement."

NISHIKAWA, M., NAGAI, T. and TSUTSUMI, H.

Osaka University, Kanto Special Steel Works

Proceedings of the International Conference on Martensitic Transformations. ICOMAT-86, Nara, Japan, 26–30 Aug. 1986

Publ.: The Japan Institute of Metals, Aoba Aramaki, Sendai 980, Japan, 1987

"Several Considerations on Design of SMA Actuator."
HIROSE, S., IKUTRA, K., TSUKAMOTO, M., SATO, K. and UMETANI, Y.
Tokyo Institute of Technology
Proceedings of the International Conference on Martensitic Transformations. ICOMAT-86, Nara, Japan, 26–30 Aug. 1986
Publ.: The Japan Institute of Metals, Aoba Aramaki, Sendai 980, Japan, 1987

"Characteristics of SMA (Shape Memory Alloy) Coupler Manufactured with DI Method."
HANAKI, K., KATO, K., IKUNO, H. and IKEDA, K.
Tokyo Institute of Technology
Proceedings of the International Conference on Martensitic Transformations. ICOMAT-86, Nara, Japan, 26–30 Aug. 1986
Publ.: The Japan Institute of Metals, Aoba Aramaki, Sendai 980, Japan, 1987

"Heat Conduction in Ni–Ti Shape Memory Allow in the Presence of Phase Transformation under Stress."
NAKAI, K., OKUDA, T. and TANAKA, H.
Sharp
Proceedings of the International Conference on Martensitic Transformations. ICOMAT-86, Nara, Japan, 26–30 Aug. 1986
Publ.: The Japan Institute of Metals, Aoba Aramaki, Sendai 980, Japan, 1987

"Martensitic Transformation of Two Phase Microstructures."
HORNBOGEN, E.
Ruhr-Universitat Bochum
Proceedings of the International Conference on Martensitic Transformations. ICOMAT-86, Nara, Japan, 26–30 Aug. 1986
Publ.: The Japan Institute of Metals, Aoba Aramaki, Sendai 980, Japan, 1987

"Domain Walls in Shape Memory Alloys as Shock Waves."
FALK, F. and SEIBEL, R.
Universitat-GH-Paderborn, Technische Universitat Berlin
Proceedings of the International Conference on Martensitic Transformations. ICOMAT-86, Nara, Japan, 26–30 Aug. 1986
Publ.: The Japan Institute of Metals, Aoba Aramaki, Sendai 980, Japan, 1987
70–81

"Production and Shape Memory Effect of Nickel Titan."
EHRENSTEIN, H.
ROKA
Proceedings of the International Conference on Martensitic Transformations. ICOMAT-86, Nara, Japan, 26–30 Aug. 1986
Publ.: The Japan Institute of Metals, Aoba Aramaki, Sendai 980, Japan, 1987

"Development of Novel "Adaptive/Smart Composite Material Utilizing Nitinol Fibers"
ROGERS, C. A. and ROBERTSHAW, H. H.
Proceedings of Third Technical Conference on Composite Materials, AM. Soc. Comp. 25–29 Sept. 1988, Seattle, WA, USA
Publ.: Technomic Publishing Company

"A Practical Shape Memory Electromechanical Actuator."
YAEGER, J. R.
ISATA 84 Proceedings, International Symposium on Automation Technology and Automation, Milan, Italy, 24–28 Sept., 1984, Vol. 1, pp. 633–642

"Application of Shape Memory Alloys to Robotic Actuators."
HASHIMOTO, ET AL.
Journal of Robotic Systems, Vol. 2, No. 1 (Spring, 1985) pp. 3–25.

"Shape Memory Alloy Application for Sequential Operation Control."
MIWA, Y.
System and Control (Japan), Vol. 29, No. 5 (May 1985), pp. 303–310

"Shape Memory Alloy Reinforced Composites."
ROGERS, C. A. and ROBERTSHAW, H. H.
Engineering Science Preprints 25, ESP25.88027, Society of Engineering Sciences, Inc., June 20–22, 1988

"Nickel-Base Alloys."
BUEHLER, W. J. and WILEY, R. C.
U.S. Patent 3,174,851, March 23, 1965

"The Shape Memory ('Marmen') Effect in Alloys." *Metal Science J.*, Vol. 6, 1972, p. 175

Shape Memory Effects in Alloys
PERKINS, J., ED.
Plenum Press, New York, 1975

"A Source Manual for Information on Nitinol and NiTi."
GOLDSTEIN, D.
Naval Surface Weapons Center, Silver Spring, Maryland, Report NSWC/WOL TR 78–26, 1978

"Shape Memory Alloys."
SCHETKY, L.
Scientific American, Vol. 241, 1979, p. 74

"55-Nitinol—The Alloy with a Memory—Its Physical Metallurgy, Properties, and Applications."
JACKSON, C. M., WAGNER, H. J. and WASILEWSKI, R. J.
NASA-SP-5110, 1972, 91 p.

"Thermoelasticity, Pseudoelasticity and the Shape-Memory Effects Associated with Martensitic Transformations."
DELAEY, R. V., TAS KRISHNAN, H. and WARLIMONT, H.
Journal of Material Science, Vol. 9, 1974, pp. 1521–1545

"Crystallographic Similarities in Shape-Memory Martensites."
SABURI, T. and WAYMAN, C. M.
Acta Metallurgica, Vol. 27, 1979, p. 979

"Nitinol Characterization Study."
CROSS, W. B., KARIOTIS, A. H. and STIMLER, F. J.
NASA CR-1433, Sept. 1969

"Development of a Novel Smart Material."
ROGERS, C. A. and ROBERTSHAW, H. H.
Proceedings of the ASME 1988 Winter Annual Meeting, November 28–December 2, 1988 (in publication)

"Active Modal Modification of Quasi-Isotropic Shape Memory Alloy Reinforced Plates."
ROGERS, C. A. and LIANG, CHEN
Proceedings of the 30th Structures, Structural Dynamics and Materials Conference, Mobile, AL, April 3–5, 1989 (submitted for review)

"A-Proof-of-Concept Experimental Investigation on a Four-Bar Linkage with a Smart Rocker Link Featuring a Shape Memory Alloy."
HANSKNECHT, B., TRAHAN, D., GANDHI, M. V., and THOMPSON, B. S.
To be presented at the 1st National Mechanisms Conference, Cincinnati, Ohio, Nov. 1989

"Dynamic Control Concepts Using Shape Memory Alloy Reinforced Plates."
ROGERS, C. A.
Proceedings of the U.S. Army Research Office Workshop, September 15–16, 1988, Blacksburg, Virginia

"Optical Fiber Sensors and Single Processing for Smart Materials and Structures Applications."
CLAUS, R. O.
Proceedings of the U.S. Army Research Office Workshop, September 15–16, 1988, Blacksburg, Virginia

3. Piezo-Electric Actuators

"Pulse Mode Operation for Piezo-Electric Ceramic Actuator."
TAKAHASHI, S., YANO, T., HORI, S. and SATO, E.
NEC Corp., Kawasaki, Japan
Jpn. Appl. Phys. Suppl. (Japan) Vol. 26, Suppl. 26-2 180-2 1987
6th Meeting on Ferroelectric Materials and Their Applications (FMA-6) 3–5 June 1987 Kyoto, Japan

"Piezo-Electric Materials for Low Hysteresis Actuators."

TANUMA, C., YOSHIDA, S. and YAMASHITA, Y.
Res. and Dev. Center, Toshiba Corp., Kawasaki, Japan

Jpn. J. Appl. Phys. Suppl. (Japan) Vol. 26, Suppl. 26-2, pp. 174–176, 1987

6th Meeting on Ferroelectric Materials and Their Applications (FMA-6) 3–5 June 1987, Kyoto, Japan

"On the Responsibility of Piezo-Electric Actuator."

FUJII, H. and OHNISHI, Y.
Sumitomo Special Metals Co. Ltd., Osaka, Japan

Jpn. J. Appl. Phys. Suppl. (Japan) Vol. 26, Suppl. 26-2, pp. 177–179, 1987

6th Meeting on Ferroelectric Materials and Their Applications (FMA-6) 3–5 June 1987, Kyoto, Japan

"Acoustic Emission in Piezo-Electric/Electrostructive Actuators."

UCHINO, K., HIROSE, T. and VARAPRASAD, A. M.
Dept. of Phys., Sophia Univ., Tokyo, Japan

Jpn. J. Appl. Phys. Suppl. (Japan) Vol. 26, Suppl. 26-2, pp. 167–170, 1987

6th Meeting on Ferroelectric Materials and Their Applications (FMA-6) 3–5 June 1987, Kyoto, Japan

"Use of Piezo-Electric Actuators as Elements of Intelligent Structures."

CRAWLEY, E. F. and de LUIS, J.
MIT, Cambridge, MA, USA

AIAA J. (USA) Vol. 25, No. 10, pp. 1373–1385, Oct. 1987

Presents the analytic and experimental development of piezo-electric actuators as elements of intelligent structures, i.e., structures with highly distributed actuators, sensors, and processing networks. Static and dynamic analytic models are derived for segmented piezoelectric actuators that are either bonded to an elastic substructure or embedded in a laminated composite. These models lead to the ability to predict, a priori, the response of the structural member to a command voltage applied to the piezo-electric device and give guidance as to the optimal location for actuator placement. A scaling analysis is performed to demonstrate that the effectiveness of piezo-electric actuators is independent of the size of the structure and to evaluate various piezo-electric materials based on their specimens of cantilevered beams were constructed: an aluminum beam with surface-bonded actuators, a glass-epoxy beam with embedded actuators. The actuators were used to excite steady-state resonant vibrations in the cantilevered beams. The response of the specimens compared well with those predicted by the analytical models. Static tensile tests performed on glass-epoxy laminates indicated that the embedded actuator reduced the ultimate strength of the laminate by 20%, while not significantly affecting the global elastic modulus of the specimen.

"Piezo-Electric Actuators and Their Applications."

TAKAHASHI, S.
NEC Corp., Kawasaki, Japan

J. Inst. Electron. Inf. Commun. Eng. (Japan) Vol. 70, No. 3, pp. 295–297, March 1987

"Piezo-Electric Bimodal Actuator Control Using a Kalman Filter."

TAMARU, N.
NTT Elect. Commun. Labs., Musashino, Japan

Trans. Soc. Instrum. Control Eng. (Japan) Vol. 23, No. 8, pp. 873–875, Aug. 1987

"Application of Piezo-Electric Polymer to Actuator."

SEO, I.

J. Jpn. Soc. Precis. Eng. (Japan) Vol. 52, No. 5, pp. 689–691, May 1987

"Actuator Applications of Piezo-Electric/Electro-Strictive Ceramics."

UCHINO, K.

J. Jpn. Soc. Precis. Eng. (Japan) Vol. 53, No. 5, pp. 686–688, May 1987

"Multilayer Piezo-Electric Actuator"
NISHIZAWA, T., SANO, M. and TAGAMI, S.
NEC TECH. J. (Japan) Vol. 40, No. 5, pp. 118–122, May 1987

"Multilayer Piezo-Electric Ceramic Actuators."
WERSING, W., SCHNOLLER, M. and WAHL, H.
Siemens, AG, Munich, Germany
Sponsor: IEEE; Army Res. Office; Office Naval Res. et al.
ISAF '86. Proceedings of the Sixth IEEE International Symposium on 8–11 June 1986, Bethlehem, PA, USA

"A Variational Formulation for the Finite Element Analysis of Linkage and Robotic Systems Featuring Smart Links Incorporating Piezo-Electric Materials."
THOMPSON, B. S. and GANDHI, M. V.
To be presented at the 1st National Mechanisms Conference, Cincinnati, Ohio, Nov. 1989.

4. Piezo-Electric Sensors

"Distributed Piezo-Electric-Polymer Active Vibration Control of a Cantilever Beam."
BAILEY, T. and HUBBARD, J. E.
Journal of Guidance, Control and Dynamics, Vol. 8, No. 5 (Sept.–Oct., 1985), pp. 605–611

"Distributed Structural Control Using Multilayered Piezo-Electric Actuators."
CUDNEY, H. H., OSHMAN, Y. and INMAN, D. J.
29th Structures, Structural Dynamics, and Materials Conference, April, 1988, Williamsburg, VA

"Distributed Parameter Actuators for Structural Control."
CUDNEY, H. H., INMAN, D. J. and OSHMAN, Y.
Proceedings of the 1989 American Control Conference

"Distributed Structural Control Using Multilayered Piezoelectric Actuators."
CUDNEY, H. H., INMAN, D. J. and OSHMAN, Y.
Proceedings of the 7th VPI & SU/AIAA Symposium on Dynamics and Control of Large Structures, 1987

"Accurate Measurement of the Voltage Microdisplacement Correlation of Piezo-Electric Sensor with a Fiber Optic Interferometer."
DONG TAI-HUO, PAO CHENG-KANG and KIN DAN
Dept. of Opt. Instrum., Zhejiang Univ., China
Sponsor: SPIE; IEEE; Comite Belge Opt. et al.
Proc. SPIE—Int. Soc. Opt. Eng. (USA) Vol. 798, pp. 304–306, 1987
Fiber Optic Sensors II, 31 March–3 April 1987. The Hague, Netherlands

"Equivalent Circuit for Piezo-Electric Vibratory Gyro of an Angular Rate Sensor."
KONNO, M., SUGAWARA, S., OYAMA, S., and NAKAMURA, H.
Fac. of Eng., Yamagata Univ., Yonezawa, Japan
Trans. Inst. Electron. Inf. Commun. Eng. A (Japan) Vol. J70A, No. 11, pp. 1724–1727, Nov. 1987

"Piezo-Electric Polymer Pressure Sensors."
LEAVER, P., CUNNINGHAM, M. J. and JONES, B. E.
Dept. of Electr. Eng., Manchester, Univ., U.K.
Sens. Actuators (Switzerland) Vol. 12, No. 3, pp. 225–233, Oct. 1987
2nd International Meeting on Chemical Sensors, 7–10 July 1986, Bordeaux, France

"An Extended Acoustic Sensor Utilising a New Piezo-Electric Cable."
FOX, D. R., ATKINSON, E. B. and PENNECK, R. J.
Raychem Ltd., Swindon, England

McAVOY, B. R. (EDITORS)
Sponsor: IEEE, Andersen Labs., Hewlett-Packard, Phonon Corp., Westinghouse et al.
IEEE 1986 Ultrasonics Symposium Proceedings, 17–19 Nov. 1986, Williamsburg, VA, USA
Publ.: IEEE, New York, USA
2 vol. 1130 pp.

"Robot Gripper Control System Demonstrating PVDF Piezo-Electric Sensors."

BARSKY, M. F., LINDER, D. K. and CLAUS, R. O.

Dept. of Electr. Eng., Virginia Polytech. Inst. and State Univ., Blacksburg, VA, USA

McAvoy, B. R. (Editors)

Sponsor: IEEE Andersen Labs., Hewlett-Packard, Phonon Corp., Westinghouse et al.

IEEE 1986 Ultrasonics Symposium Proceedings (Cat. No. 86CH2375-4) 545-8, Vol. 1, 1986

17–19 Nov. 1986, Williamsburg, VA, USA

Publ.: IEEE, New York, USA

"The Method to Increase Reliability of Piezo-Electric Angular Rate Sensors."

ZHANG, FU ZUE

Sensor Electron. Inst., Beijing Inf. Technol. Inst., China

Reliab. Eng. (GB), Vol. 19, No. 1, pp. 15–21, 1987

"Relative Humidity Measurements Using a Coated Piezo-Electric Quartz Crystal Sensor."

RANDIN, J. P. and ZULLIG, F.

ASULAB Inc., Neuchatel, Switzerland

Sens. and Actuators (Switzerland) Vol. 11, No. 4, pp. 319–28, May–June 1987

"Samariaum Doped Potassium-Niobate—A New Piezo-Electric Sensor."

MOTGHARE, S. J., KOLTE, A. M., DHULEY, C. V. and KARPATAL, A. G.

Dept. of Appl. Phys., Visvesvaraya Regional Coll. of Eng., Nagpur, India

Sponsor: IEEE, Army Res. Office, Office Naval Res. et al.

ISAF '86. Proceedings of the Sixth IEEE International Symposium on Applications of Ferroelectrics (Cat. No. 86CH2358-0) 472–5, 1986

8–11 June 1986, Bethlehem, PA, USA

Publ.: IEEE, New York, USA

"Piezoelectric Vibration Sensor with Thermal Compensation."

ZUSMAN, G.V., NIKULIN, A. A. and PETROVICH, V. I.

Izmer, Tekh. (USSR) Vol. 28, No. 10, pp. 22–23, Oct. 1985

Trans in: *Meas. Tech. (USA)* Vol. 28, No. 10, pp. 843–844, Oct. 1985

"Active Damping of a Robotic Gripper PVDF Piezo-Electric Sensor."

BARSKY, M. F., LINDNER, D. K. and CLAUS, R. O.

Dept. of Electr. Eng., Virginia Polytech. Inst. and State Univ., Blacksburg, VA, USA

Sponsor: IEEE; Univ. Tennessee

Proceedings of the Eighteenth Southeastern Symposium on System Theory

7–8 April 1986, Knoxville, TN, USA

Publ.: IEEE Comput. Soc. Press, Washington, DC, USA

"Composite Piezo-Electric Sensors."

SAFARI, A., SA-GONG, G., GINIEWICZ, J. and NEWNHAM, R. E.

Mater. Res. Lab., Pennsylvania State Univ., University Park, PA, USA

Tressler, R. E., Mesing, G. L., Pantano, C. G., Newnham, R. E. (Editors)

"Tailoring Multiphase and Composite Ceramics."

Proceedings of the Twenty-First University Conference on Ceramic Science, pp. 445–454, 1986

17–19 July 1985, University Park, PA, USA

Publ.: Plenum, New York, USA

"Chemical Piezo-Electric Sensor and Sensor Array Characterization."

CAREY, W. P. and KOWALSKI, B. R.

Dept. of Chem., Washington Univ., Seattle, WA, USA

Anal. Chem. (USA) Vol. 58, No. 14, pp. 3077–3084, Dec. 1986

"Piezoelectric Plastics Promise New Sensors."

CARLISLE, B. H.

Mach. Des. (USA) Vol. 58, No. 25, 105–110, 23 Oct. 1986

"A Piezo-Electric Film Sensors with Uniformly Expanded Surface to Detect Tactile Information for Robotic End-Effectors."

NAKAMURA, Y., HANAFUSA, H. and UENO, N.

Proceedings of the '85 International Conference on Advanced Robotics, Tokyo, Japan, 9–10, Sept. 1985, pp. 137–144

"Piezo-Electric Ceramic Actuator and Its Applications."

TAKAHASHI, S.

Oyo Buturi (Japan) Vol. 54, No. 6, pp. 587–588 (June 1985). In Japanese.

"Vibration Analysis of Piezo-Electric Actuators."

TANOSHIMA, K., ARAKI, T. and TSUKADA, M.

IEEE 1984 Ultrasonics Symposium Proceedings. Dallas, Texas, 14–16 Nov. 1984, Vol. 2, pp. 882–887

"Piezo-Electric Driven Turntable with High Positioning Accuracy."

TOJO, T., and SUGIHARA, K.

Bulletin of the Japanese Society of Precision Engineers, Vol. 19, No. 2 (June 1985), pp. 135–137

"Nonlinear Control of a Distributed System: Simulation and Experimental Results." *ASME Journal of Dynamic Systems, Measurements, and Control*, Vol. 109, 1987, pp. 133–139

"Smart Components for Structural Vibration Control."

MILLER, S. E. and HUBBARD, JR., J. E.

Proceedings of the 1988 American Control Conference, Atlanta, GA, Vol. 3, 1988, pp. 1897–1902.

"Development of Piezo-Electric Technology for Applications in Control of Intelligent Structures."

CRAWLEY, E. F., de LUIS, J., HAGOOD, N. W. and ANDERSON, E. H.

Proceedings of the 1988 American Control Conference, Atlanta, GA, Vol. 3, 1988, pp. 1897–1902

"Piezo-Ceramic Devices and PVDF Films as Sensors and Actuators for Intelligent Structures."

HANAGUD, S.

Proceedings of the 1988 U.S. Army Research Office' Workshop, Blacksburg, Virginia, September 1988

"Materials Issues for Smart Structures."

WILKES, G. L.

Proceedings of the 1988 U.S. Army Research Office Workshop, Blacksburg, Virginia, September 1988

"Smart Ceramics."

NEWNHAM, R. E.

Proceedings of the 1988 U.S. Army Research Office Workshop, Blacksburg, Virginia, September 1988

5. Fiber Optic Sensors

"Embedded Optical Fiber Sensors for Intelligent Aerospace Structures."

WIENCKO, J. A., CLAUS, R. O. and ROGERS, R. E.

Dept. of Electr. Eng., Virginia Polytech., Blacksburg, VA, USA

Sponsor: Sensors Magazine

Proceedings of SENSORS EXPO, 15–17 Sept. 1987, Detroit, MI, USA, pp. 257–262, 1987

Publ.: Helmers Publishing, Peterborough, NH, USA, xiv + 487 pp.

Optical fiber sensors embedded within aerospace material structures offer a method for the internal monitoring of the structure from its manufacture to its repair or replacement. The authors consider the application of such sensors in intelligent structures of advanced composite materials. Optical fibers have been embedded between graphite-epoxy and PEEK prepreg laminae and used to monitor internal mechanical properties during temperature and pressure cure cycle processing. The same sensor fibers, now integral parts of the laminate, have been used to measure feedback control system parameters such as strain and vibration during simulated material use lifetimes. The fibers have also been used to nondestructively determine the onset of material failure via the detection of acoustic emission signatures in loaded material structures. Methods for the embedding of optical fibers in advanced composites as well as methods for sensing and signal processing are reviewed.

"Line Independence of Optical Fibre Sensors."

KIST, R.

Fraunhofer Inst. fur Phys., Freiburg, West Germany

Technica (Switzerland) Vol. 37, No. 4, pp. 13–16, 2 March 1988

"Coherence Multiplexing of Remote Fibre Optic Fabry-Perot Sensing System

FARAHI, F., NEWSON, T. P., JONES, J. D. C. and JACKSON, D. A.

Phys. Lab., Kent Univ., Canterbury, UK

Opt. Commun. (Netherlands) Vol. 65, No. 5, pp. 319–321, 1 March 1988

"Optical Fibre Sensing Using GRIN Rod Lenses."

CUSWORTH, D. S. and SENIOR, J. M.

Dept. of Electr. and Electron. Eng., Fac. of Sci. and Eng., Manchester Polytech., UK

Int. J. Opt. Sens. (UK) Vol. 2, No. 5–6, pp. 421–436, Oct.–Dec. 1987

"Assembling Technique of Microcapsules for Optical Fibre Sensors."

BRENCI, M., CONFORTI, G., FALCIAI, R., MIGNANI, A. G. and GIRONI, G.

IROE-CNR Firenze, Italy

Int. J. Opt. Sens. (UK) Vol. 2, No. 5–6, pp. 341–347, Oct.–Dec. 1987

"Magnetic Field Sensitivity Analysis of an Optical Fibre with Metallic Glass Strips."

ZHIPENG ZHAND, QIN ZHANG

Dept. of Electr. Power Eng., Huazhong Univ. of Sci. and Technol., Wuhan, China

Int. J. Opt. Sens (UK) Vol. 2, No. 5–6, pp. 341–347, Oct.–Dec. 1987

"Fibre Optic Mechanical Sensors for Aerospace Applications."

BATCHELLOR, C. R., DAKIN, J. P. and PEARCE, D. A. J.

Plessey Res. Roke Manor Ltd., Romsey, UK

Int. J. Opt. Sens. (UK) Vol. 2, No. 5–6, pp. 407–411, Oct.–Dec. 1987

"A Novel Distributed Optical Fibre Sensing System Enabling Location of Disturbances in a Sagnac Loop Interferometer."

DAKIN, J. P., PEARCE, D. A. J., STRONG, A. P. and WADE, C. A.

Plessey Res. Roke Manor Ltd., Romsey, UK

Int. J. Opt. Sens. (UK) Vol. 2, No. 5–6, pp. 403–406, Oct.–Dec. 1987

"Optical Fibre Sensor."

TSUTSUI, T. and YAMAMOTO, S.

J. Jpn. Soc. Precis. Eng. (Japan) Vol. 53, No. 12, pp. 1847–1851, Dec. 1987

"Development of Intelligent Optical Fiber Sensors."

HATTORI, H.

Instrumentation (Japan) Vol. 30, No. 12, pp. 25–28, Dec. 1987

"High-Resolution Correlation OTDR for Distributed Fiber Optic Sensors and Mobile Cabling."

BERNARD, J.-J. and DEPRESLES, E.

Labs. de Marcoussis, CGE Res. Center, France

Fiber Integr. Opt. (USA) Vol. 7, No. 2, pp. 79–84, 1988

"Fiber Optic Gyroscope: An Advanced Rotation Rate Sensor."

AUCH, W., SCHLEMPER, E. and WENZEL, W.

Standard Elektrik Lorenz AG, Stuttgart, West Germany

Electr. Commun. (UK), Vol. 61, No. 4, pp. 372–378, 1987

"Multiplexed Fibre Optic Interferometric Sensing System: Combined Frequency and Time Division."

FARAHI, F., JONES, J. D. C. and JACKSON, D. A.

Phys. Lab., Kent Univ., Canterbury, UK

Electron. Lett. (UK) Vol. 24, No. 7, pp. 409–410, 31 March 1988

"Mode-Mode Interference Effects in Axially Strained Few-Mode Optical Fibers."
SHANKARANARAYANAN, N. K., SRINIVAS, K. T. and CLAUS, R. O.
Fiber and Electro Optics Res. Center, Virginia Tech., Blacksburg, VA, USA
Sponsor: SPIE
Proc.: SPIE—Int. Soc. Opt. Eng. (USA) Vol. 838, pp. 385–388, 1988

"Optical Performance of a Fused Silica Flexible Imaging Device."
GERACI, D. and FRANCAVILLA, S.
Galileo Electro-Opt. Corp., Sturbridge, MA, USA
Sponsor: SPIE
Proc. SPIE—Int. Soc. Opt. Eng. (USA) Vol. 838, pp. 379–384, 1988
Fiber Optic and Laser Sensors V, 17–19 Aug. 1987, San Diego, CA, USA

"Long-Lead Cable Effects on Interferometric Sensors."
FREIDAH, J. T., WAGONER, R. E., CASH, T. J. and SAFAR, N. H.
McDonnell Douglas Astronaut Co., Huntington Beach, CA, USA
Sponsor: SPIE
Proc. SPIE—Int. Soc. Opt. Eng. (USA) Vol. 838, pp. 372–378, 1988
Fiber Optic and Laser Sensors V, 17–19 Aug. 1987, San Diego, CA, USA

"Single-Mode Fiber Pseudo-Depolarizer."
KERSEY, A. D., DANDRIDGE, A. and MARRONE, M. J.
Naval Res. Lab., Washington, DC, USA
Sponsor: SPIE
Proc. SPIE—Int. Soc. Opt. Eng. (USA) Vol. 838, pp. 360–364, 1988
Fiber Optic and Laser Sensors V, 17–19 Aug. 1987, San Diego, CA, USA

"Silicone-Clad Fiber Sensor for Detecting of Water Penetration into Optical Cables."
BUBNOV, M. M., ABRAMOV, A. A., BOGATYRJOV, V. A. and DIANOV, E. M.
Gen. Phys. Inst., Acad. of Sci., Moscow, USSR
Sponsor: SPIE
Proc. SPIE—Int. Soc. Opt. Eng. (USA) Vol. 838, pp. 344–346, 1988
CODEN: PSISDG ISSN: 0277-786X
Fiber Optic and Laser Sensors V, 17–19 Aug. 1987, San Diego, CA, USA

"Passive Stabilization of a Fiber Optic Nonlinear Sensor Using a Synchronous Sampling Scheme."
KOO, K. P., BUCHOLTZ, F. and DANDRIDGE, A.
Naval Res. Lab., Washington, DC, USA
Sponsor: SPIE
Proc. SPIE—Int. Soc. Opt. Eng. (USA) Vol. 838, 340–393, 1988
CODEN: PSISDG ISSN: 0277-786X
Fiber Optic and Laser Sensors V, 17–19 Aug. 1987, San Diego, CA, USA

"Elliptical Core Two-Mode Fiber Strain Gauge."
BLAKE, J. N., HUANG, S. Y. and KIM, B. Y.
Edward L. Ginzton Lab., Stanford Univ., CA, USA
Sponsor: SPIE
Proc. SPIE—Int. Soc. Opt. Eng. (USA) Vol. 838, 332–339, 1988
Fiber Optic and Laser Sensors V, 17–19 Aug. 1987, San Diego, CA, USA

"Fiber Optic Spectrometer."
HIGLEY, S. E., UDD, E., MICHAL, R. J., THERIAULT, J. P. and JOLIN, D. A.
McDonnell Douglas Astronaut Co., Huntington Beach, CA, USA
Sponsor: SPIE
Proc. SPIE—Int. Soc. Opt. Eng. (USA) Vol. 838, pp. 318–324, 1988

"Porous Glass Fibers for High Sensitivity Chemical and Biomedical Sensors."
SHAHRIARI, M. R., SIGEL, G. H. and QUAN ZHOU
Rutgers Univ., Piscataway, NJ, USA
Sponsor: SPIE
Proc. SPIE—Int. Soc. Opt. Eng. (USA) Vol. 838, pp. 348–352, 1988
CODEN: PSISDG ISSN: 0277-786X
Fiber Optic and Laser Sensors V, 17–19 Aug. 1987, San Diego, CA, USA

"Mode-Mode Interference Effects in Axially Strained Few-Mode Optical Fibers."

SHANKARANARAYANAN, N. K., SRINIVAS, K. T. and CLAUS, R. O.

Fiber & Electro Optics Res. Center, Virginia Tech., Blacksburg, VA, USA

Sponsor: SPIE

Proc. SPIE—*Int. Soc. Opt. Eng. (USA)* Vol. 838, pp. 385–388, 1988

Fiber Optic and Laser Sensors V, 17–19 Aug. 1987, San Diego, CA, USA

"Optical Fiber Modal Domain Detection of Stress Waves."

SHANKARANARAYANAN, N. K., BENNETT, K. D. and CLAUS, R. O.

Virginia Tech., Blacksburg, VA, USA

McAvoy, B. R. (Editor)

Sponsor: IEEE, Andersen Labs., Hewlett-Packard, Phonon Corp., Westinghouse, et al.

IEEE 1986 Ultrasonics Symposium Proceedings (Cat. No. 86CH2375-4) 1071-4, Vol. 2, 1986

17–19 Nov. 1986, Williamsburg, VA, USA

Publ.: IEEE, New York, USA

"Single Mode Optical Fiber Vibration Sensor."

FLAX, A., PENNINGTON, C. and CLAUS, R. O.

Dept. of Electr. Eng., Virginia Tech., Blacksburg, VA, USA

Conference Proceedings. IEEE SOUTHEASTOCON '87 (Cat. No. 87CH2398-6), Vol. 2, pp. 553–555, 1987

5–8 April 1987, Tampa, FL, USA

Publ.: IEEE, New York, USA

"Acoustic Fiber Waveguide Devices."

MATTHEWS, A. L., MURPHY, K., SAFAAI-JAZI, A. and CLAUS, R. O.

Dept. of Electr. Eng., Virginia Tech., Blacksburg, VA, USA

Conference Proceedings. IEEE SOUTHEASTOCON '87 (Cat. No. 87CH2398-6), Vol. 2, pp. 533–541, 1987

5–8 April 1987, Tampa, FL, USA

Publ.: IEEE, New York, USA

"Fiber Optic Composite Impact Monitor."

KUHLMAN, R., DUNCAN, B. and CLAUS, R. O.

Dept. of Electr. Eng., Virginia Polytech. Inst. & State Univ., Blacksburg, VA, USA

Conference Proceedings. IEEE SOUTHEASTOCON '87 (Cat. No. 87CH2398-6), Vol. 2, pp. 414–417, 1987

5–8 April 1987, Tampa, FL, USA

Publ.: IEEE, New York, USA

"Nondestructive Testing of Composite Materials by OTDR in Imbedded Optical Fibers."

CLAUS, R. O., JACKSON, B. S. and BENNETT, K. D.

Dept. of Electr. Eng., Virginia Polytech. Inst., State Univ., Blacksburg, VA, USA

Sponsor: SPIE

Proc. SPIE Int. Soc. Opt. Eng. (USA) Vol. 566, pp. 243–248, 1985

"Analysis of Composite Structures Using Optical Fiber Modal Sensing Techniques."

BENNETT, K. D. and CLAUS, R. O.

Dept. of Electr. Eng., Virginia Polytech. Inst., State Univ., Blacksburg, VA, USA

Sponsor: IEEE

Conference Proceedings IEEE SOUTHEASTCON '86 (Cat. No. 86CH2306-9), pp. 95–98, 1986

23–25 March 1986, Richmond, VA, USA

Publ.: IEEE, New York, USA

"Concentric Core Optical Fiber Strain Sensor."

SHIH, J. C. and CLAUS, R. O.

Dept. of Electr. Eng., Virginia Polytech. Inst., State Univ., Blacksburg, VA, USA

Sponsor: IEEE

Conference Proceedings IEEE SOUTHEASTOCON '86 (Cat. No. 86CH2306-9) pp. 91–94, 1986

23–25 March 1986, Richmond, VA, USA

Publ.: IEEE, New York, USA

"Monitoring of Strain in Layered Media Using Clad Rod Acoustic Waveguides."
BENNETT, K. D., HANNA, S. J. and CLAUS, R. O.
Dept. of Electr. Eng., Virginia Polytech. Inst., State Univ., Blacksburg, VA, USA
McAvoy, B. R. (Editor)
Sponsor: IEEE, Allied Corp., Andersen Labs., Crystal Technol. et al.
IEEE 1985 Ultrasonics Symposium Proceedings (Cat. No. 85CH2209-5), Vol. 2, pp. 1064–1067, 1985
16–18 Oct. 1985, San Francisco, CA, USA
Publ.: IEEE, New York, USA

"Nondestructive Evaluation of Composites by Optical Time Domain Reflectometry in Embedded Optical Fibers."
CLAUS, R. O., JACKSON, B. S. and MAY, R. G.
Dept. of Electr. Eng., Virginia Polytech. Inst. and State Univ., Blacksburg, VA, USA
Sponsor: IEEE
Conference Proceedings IEEE SOUTHEASTCON '85 (Cat. No. 85CH2161-8) 241–5, 1985
31 March–3 April 1985, Raleigh, NC, USA
Publ.: IEEE, New York, USA

"Single Mode Acoustic Fiber Waveguide."
JACKSON, B. S., MAY, R. G. and CLAUS, R. O.
Dept. of Electr. Eng., Virginia Polytech. Inst. and State Univ., Blacksburg, VA, USA
Sponsor: IEEE
Conference Proceedings of IEEE SOUTHEASTCON '84, pp. 226–230, 1984
8–11 April 1984, Louisville, KY, USA
Publ.: IEEE, New York, USA

"Optical Fiber Sensor Application in a Simple Robotic Gripper."
O'CONNER, W. T. and CLAUS, R. O.
Dept. of Electr. Eng., Virginia Polytech. Inst. and State Univ., Blacksburg, VA, USA
Sponsor: IEEE
Conference Proceedings of IEEE SOUTHEASTCON '84, pp. 295–297, 1984
8–11 April 1984, Louisville, KY, USA
Publ.: IEEE, New York, USA

"Calculation of Ultrasonic Field Propagation in Fluids Using Finite Difference Techniques."
DOCKERY, G. D. and CLAUS, R. O.
Dept. of Electr. Eng., Virginia Polytech. Inst. and State Univ., Blacksburg, VA, USA
McAvoy, B. R. (Editor)
Sponsor: IEEE
1983 Ultrasonics Symposium Proceedings, Vol. 1, pp. 508–511, 1983
31 Oct.–2 Nov., 1983, Atlanta, GA, USA
Publ.: IEEE, New York, USA

"Radiation Pattern Model for Partially Electroded Piezoelectric Transducers with Ring Electrodes."
GRAY, J. W., O'CONNER, W. T., BURRIER, R. A. and CLAUS, R. O.
Dept. of Electr. Eng., Virginia Polytech. Inst. and State Univ., Blacksburg, VA, USA
McAvoy, B. R. (Editor)
Sponsor.: IEEE
1983 Ultrasonics Symposium Proceedings, Vol. 1, pp. 504–507, 1983
31 Oct.–2 Nov., 1983, Atlanta, GA, USA
Publ.: IEEE, New York, USA

"Measurement of Ultrasonic Fields in Transparent Media Using a Scanning Differential Interferometer."
DOCKERY, G. D. and CLAUS, R. O.
Dept. of Electr. Eng., Virginia Polytech. Inst. and State Univ., Blacksburg, VA, USA
McAvoy, B. R. (Editor)
Sponsor: IEEE
1983 Ultrasonics Symposium Proceedings, Vol. 2, pp, 573–577, 1983
31 Oct.–2 Nov., 1983, Atlanta, GA, USA
Publ.: IEEE, New York, USA

"Interferometric Techniques Using Embedded Optical Fibers for the Quantitative NDE of Composites."
WADE, J. C. and CLAUS, R. O.
Dept. of Electr. Eng., Virginia Polytech. Inst. and State Univ., Blacksburg, VA, USA
Thompson, D. O., Chimenti, D. E. (Editors)
Review of Progress in Quantitative Nondestructive Evaluation, Vol. 2B, pp. 1731–1738, 1983
1–6 Aug. 1982, San Diego, CA, USA

"Application of Optical Fibers to Wide-Band Differential Interferometry."

GARG, A. O. and CLAUS, R. O.

Virginia Polytech. Inst. and State Univ., Blacksburg, VA, USA

Mater. Eval. (USA) Vol. 41, No. 1, pp. 106–109, Jan. 1983

"DC Calibration of the Strain Sensitivity of a Single Mode Optical Fiber Interferometer."

CLAUS, R. O. and CANTRELL, J. H., JR.

Dept. of Electrical Engng., Virginia Polytech. Inst. and State Univ., Blacksburg, VA, USA

Sponsor: IEEE

IEEE SOUTHEASTCON 1981 Conference Proceedings, pp. 829–831

5–8 April 1981, Huntsville, AL, USA

Publ.: IEEE, New York, USA

"Optical Actuator."

MORIKAWA, T.

J. Soc. Instrumentation and Control Engineering (Japan). (Sept. 1985) Vol. 24, No. 9, pp. 827–831

"Fiber Optic Electric Field Sensors Utilizing Electro-Absorption."

SU, S. F.

Conference Proceedings IEEE SOUTHEASTCON '85, Raleigh, NC, USA, 31 March–3 April 1985 (New York, USA IEEE 1985), pp. 241–245

"Fiber Optic Sensors—Future Light."

MAIN, R. P.

Sensor Review, (GB), Vol. 5, No. 3 (July, 1985) pp. 133–139

"So What Future Do You See in Fiber Optics?"

MANN, R.

Process Engineering (GB), Vol. 66, No. 6 (June, 1985) pp. 79–81

"Fibre Optic Sensors."

MARTINELLI, M.

Electron Oggi. (Italy) No. 4 (April 1984), pp. 115–116, 118, 120, 122, 124

"Fiber Optics for Propulsion Control Systems."

BAUMBICK, R. J.

Transactions of ASME Journal of Engines, Gas Turbines and Power. Vol. 107, No. 4, pp. 851–855 (Oct. 1985)

"DC Fibre Optic Accelerometer with Sub-μg Sensitivity."

BUCHOLTZ, F., KERSEY, A. D. and DANDRIDGE, A.

Electronic Letter (GB). Vol. 22, No. 9, pp. 451–453 (24 April 1986)

"Nondestructive Evaluation of Composites by Optical Time Domain Reflectometry in Embedded Optical Fibers."

CLAUS, R. O., JACKSON, B. S. and MAY, R. G.

Conference Proceedings IEEE SOUTHEASTCON '85, Raleigh, NC, USA, 31 March–3 April 1985 (New York, USA IEEE 1985), pp. 241–245

"Fiber Optic Vibration Sensors for Structural Control Applications."

SPILLMAN, W. B. and KLINE, B. R.

Proceedings of Damping '89, West Palm Beach, FL 1989, published by Flightdynamics Laboratory, Wright-Patterson Air Force Base, Ohio

Proceedings of the SPIE O-E/Fiber LASE Symposium, Boston, MA Sept. 6–10, 1988. This was the first major conference devoted to Fiber Optic Smart Structures and Skins.

"An Innovative Class of Revolutionary Articulating Mechanism and Robotic Systems Exploiting Smart Materials and Structures: Fiber Optics, Shape Memory Metals, Piezo-Electric Materials and Electro-Rheological Fluids."

GANDHI, M. V. and THOMPSON, B. S.

To be presented at the 1st National Mechanisms Conference, Cincinnati, Ohio, Nov. 1989.

"Optical Fiber Sensors and Signal Processing for Smart Materials and Structures Applications."

CLAUS, R. O.

Proceedings of the U.S. Army Research Office Workshop, September 15–16, Blacksburg, Virginia.

"Use of Optical Fibre for Damage and Strain Detection in Composite Materials."
S. R. WAITE, R. P. TATAM and A. JACKSON
Composites, Vol. 19, Number 6, November 1988.

"The Failure of Optical Fibres Embedded in Composite Materials."
WAITE, S. R. AND SAGE, G. N.
Composites, Vol. 19, Number 4, July 1988.

6. Smart Sensors and Actuators

"Opto-Mechanical Smart Actuator."
HOCKADAY, B. and WATERS, J.
United Technol. Res. Center, East Hartford, CT, USA
Sponsor: American Autom. Control Council
Proceedings of the 1987 American Control Conference, 735–736, Vol. 1, 1987
10–12 June 1987, Minneapolis, MN, USA
Publ.: American Autom. Control Council, Green Valley, AZ, USA
3 vol., 2161 pp.

"The Role of Pattern Recognition in VLSI Testing."
BEDROSIAN, S. D.
Moore Sch. of Electr. Eng., Pennsylvania Univ., Philadelphia, PA, USA
Sponsor: IEEE
International Test Conference 1986 Proceedings. Testing's Impact on Design and Technology (Cat. No. 86CH2339-0) pp. 750–754, 1986
8–11 Sept. 1986, Washington, DC, USA
Publ.: IEEE Comput. Soc. Press, Washington, DC, USA

"The Use of Distributed Microcomputers in the Control of Structural Dynamics Systems."
MONTGOMERY, R. C. and WILLIAMS, J. P.
NASA, Langley Res. Center, Hampton, VA, USA
Hamza, M. H. (Editor)
Mini and Microcomputers in Control, Filtering and Signal Processing. Proceedings of the ISMM International Symposium, pp. 140–144, 1985
10–12 Dec. 1984, Las Vegas, NV, USA
Publ.: Acta Press, Anaheim, CA, USA

"Valve Actuators Get Smart."
BLICKLEY, G. J.
Control Eng. (USA) Vol. 32, No. 10, pp. 81–82, Oct. 1985

"Machine Vision Provides Smart Sensing."
Robotics World (USA) Vol. 3, No. 1, pp. 42–43, Jan. 1985

"Array Sensor for Tactile Sensing in Robotic Applications."
JAYAWANT, B. V. and WATSON, J. D. M.
Sussex Univ., Brighton, UK
Sponsor: IEE
IEE Colloquium on "Solid State and Smart Sensors" (Digest No. 39) 1988
18 March 1988, London, UK
Publ.: IEE, London, UK

"Sensors, Intelligence and Networks."
BRIGNELL, J. E. and ATKINSON, J. K.
Dept. of Electron. & Comput. Sci., Southampton Univ., UK
Sponsor: IEE
IEE Colloquium on Solid State and Smart Sensors (Digest No. 39) 1988
18 March 1988, London, UK
Publ.: IEE, London, UK

"A Mark/Period Encoder/Decoder for Low Level Sensor Communications."
DOREY, A. P., BURD, N. C. and DRINALI, M.
Dept. of Eng., Lancaster Univ., UK
Sponsor: IEE
IEE Colloquium on Solid State and Smart Sensors (Digest No. 39) 1988
18 March 1988, London, UK
Publ.: IEE, London, UK

"Electrical Contacts to Solid State Ion-Selective Electrodes."

OWEN, A. E.

Dept. of Electr. Eng., Edinburgh Univ., UK

Sponsor: IEE

IEE Colloquium on Solid State and Smart Sensors (Digest No. 39) 1988

18 March 1988, London, UK

Publ.: IEE, London, UK

"Smart Sensors in Industry."

FAVENNEC, J.-M.

Electr. de France, St.-Denis, France

J. Phys. E (GB), Vol. 20, No. 9, pp. 1087–1090, Sept. 1987

"Application of Multiple Regression Methods and Automatic Testing Equipment to the Characterization of Smart Sensors."

LEGRAS, J. C., PRIEL, M. and RANSON, C.

Lab. Nat. d'Essais, Paris, France

Sens. Actuators (Switzerland) Vol. 12, No. 3, pp. 235–243, Oct. 1987

2nd International Meeting on Chemical Sensors, 7–10 July 1986, Bordeaux, France

Future-Generation Smart Field Sensor and Smart Field Network Instrumentation (Japan) Vol. 30, No. 11, pp. 52–53, Nov. 1987

"Tradeoffs in Automotive Smart Sensors System Partitioning."

ALBA, M.

Automotive Strategic Marketing, Nat. Semicond. Corp., Santa Clara, CA, USA

Sponsor: Sensors Magazine

Proceedings of SENSORS EXPO, p. 401, 1987

15–17 Sept. 1987, Detroit, MI, USA

Publ.: Helmers Publishing, Peterborough, NH, USA

"Smart Sensors and 'Dynamic Stare'."

CARSON, J. C.

Photonics Spectra (USA) Vol. 21, No. 5, pp. 158–161, May 1987

"Smart Sensors for Automotive Applications."

GIACHINO, J. M.

Ford Motor Co., Dearborn, MI, USA

Sponsor: Casa Risparmio Firenze, Province Florence, Thermotech. Assoc. Tuscany, Univ. Florence

16th ISATA: In Pursuit of Technical Excellence.

Proceedings of the 16th International Symposium on Automotive Technology and Automation, with Particular Reference to Automotive Micro Electronics, Vehicle Management Systems and Computer-Aided Testing, pp. 323–334, Vol. 1, 1987

11–15 May 1987, Florence, Italy

Publ.: Automotive Autom. (1984), Croydon, Surrey, England

"Generalized Smart Sensor Array Electronics."

TRONTELJ, J., TRONTELJ, L., KUNC, V., STIGLIC, M., SHENTON, G., ROBINSON, M., WARREN, G. and JUNGO, C.

Edvard Kardelj Univ., Trzaska, Yugoslavia;

Sponsor: IEEE

Proceedings of the IEEE 1987 Custom Integrated Circuits Conference (Cat. No. 87CH2430-7), pp. 708-711, 1987

4–7 May 1987, Portland, OR, USA

"Smart Sensors."

GIACHINO, J. M.

Div. of Electr. & Electron., Ford Motor Co., Dearborn, MI, USA

Sens. & Actuators (Switzerland) Vol. 10, No. 3–4, pp. 239–248, Nov.–Dec. 1986

"The Design of a Smart Sensor System for Real Time Remote Sensing Image Processing on Board Satellite."

ZHANG DAPENG, LI ZHONGRONG

Sponsor: SPIE

Proc. SPIE Int. Soc. Opt. Eng. (USA) Vol. 657, pp. 164–170, 1986

"Smart Sensors: When and Where?"
MIDDELHOEK, S. and HOOGERWERF, A. C.
Dept. of Electr. Eng., Delft Univ. of Technol., Netherlands
Sponsor: NSF, NIH, NBS
Sens. & Actuators (Switzerland) Vol. 8, No. 1, pp. 39–48, Sept. 1985

"Sensors, Smart Sensors and Machine Perception."
HELMERS, C. T., JR.
North Americal Technol. Inc., Peterborough, NH, USA
Sponsor: IEEE, ERA
Electro/86 and Mini/Micro Northeast. Conference Record, 20/2/1–7, 1986
13–15 May 1986, Boston, MA, USA
Publ.: Electron. Conventions Manage, Los Angeles, CA, USA

"Micros and Sensors Make Smart Transmitters."
TINHAM, B.
Control & Instrum. (GB), Vol. 18, No. 9, pp. 55, 57, Sept. 1986

"SI's Smart Silicon Sensor."
HANNEBORG, A. and OHLCKERS, P.
Elektro (Norway) Vol. 98, No. 15, pp. 8–10, 30 Sept. 1985

"Robot Tactile Sensing: A New Array Sensor."
JAYAWANT, B. V., ONORI, M. A. and WATSON, J. D. McK.
Dept. of Eng. & Appl. & Sci., Brighton Univ., England
Sponsor: IEE
IEE Colloquium on Solid State and Smart Sensors (Digest No. 54) 8/1–7, 1985
14 May 1985, London, England
Publ.: IEE, London, England

"Intelligent Transducers."
BURD, N. C. and DOREY, A. P.
Dept. of Eng., Lancaster Univ., England
J. Microcomput. Appl. (GB) Vol. 7, No. 2, pp. 87–97, April 1984

"Using Numerical Processors for Intelligent Sensor Design."
CARO, J.
Univ. Claude Bernard, Villeurbanne, France
Albertos, P., De La Puente, J. A. (Editors)
Sponsor: IFAC
Components, Instruments and Techniques for Low Cost Automation and Applications. IFAC Symposium, pp. 61–65, 1988
27–29 Nov. 1986, Valencia, Spain
Publ.: Pergamon, Oxford, UK

"Development of Intelligent Sensors Using Polymer Materials."
SEO, I.
Instrumentation (Japan) Vol. 30, No. 12, pp. 32–35, Dec. 1987

"Field Network of Intelligent Sensors."
USHIKUBO, S.
Instrumentation (Japan) Vol. 30, No. 12, pp. 40–44, Dec. 1987

VAN DER ZIJP, J., CHOUDRY, A. and PATRICK, S. L.
Center for Appl. Opt., Alabama Univ., Huntsville, AL, USA
Sponsor: SPIE
Proc. SPIE—Int. Soc. Opt. Eng. (USA) Vol. 840, pp. 164–168, 1987

"Adjustment of Intelligent Sensors with a Low-Cost Function Tester."
Elektronik (West Germany) Vol. 37, No. 6, pp. 134–136, 138
18 March 1988

"Development of Intelligent Optical Fiber Sensors."
HATTORI, H.
Instrumentation (Japan) Vol. 30, No. 12, pp. 25–28, Dec. 1987

"Intelligent Sensors: One Must First Be Convinced."
Mesures (France) Vol. 52, No. 7, pp. 61, 63, 65, 67, 4 May 1987

"Optical Proximity Sensor Systems for Intelligent Robot Hands."
ERSU, E. and KEGEL, G.
Isra Systemtech. GmbH, Darmstadt, West Germany
Adali, E., Tunali, F. (Editors)
Sponsor: IFAC, Tech. Univ. Istanbul, Bogazici Univ., Yildiz Univ.
Microcomputer Application in Process Control. Selected Papers from the IFAC Symposium, pp. 191–196, 1987
22–25 July 1986, Istanbul, Turkey
Publ.: Pergamon, Oxford, UK

"Intelligent Sensors. The Way for Performance Enhancement of Machine Controls."
WALCHER, H. and BARTOSZ, R.
Elektronik (West Germany) Vol. 36, No. 23, pp. 115–118, 121–122, 124, 126–128
13 Nov. 1987

"The Development of an Intelligent Sensor-Based Industrial Robotic System."
YUNG, H. C., ALLEN, C. R., FINCH, J. W., HARRIS, A. J., ADAMS, A. E. and FARSI, M.
Dept. of Electr. & Electron. Eng., Newcastle upon Tyne Univ., UK
Sponsor: IEEE, Univ. Tokyo
Proceedings of the International Workshop on Industrial Automation Systems: Seiken Symposium (Cat. No. 87TH0165-1) pp. 109–113, 1987
4–6 Feb. 1987, Tokyo, Japan
Publ.: IEEE, New York, NY, USA

"An Intelligent Power Sensor."
Electron. Puissance (France) No. 23, pp. 66–68, Oct. 1987

"A Microcomputer-Based Intelligent Sensor for Multiaxis Force/Torque Measurement."
LUO, R. C.
Dept. of Electr. & Comput. Eng., North Carolina State Univ., Raleigh, NC, USA
IEEE Trans. Ind. Electron. (USA) Vol. 35, No. 1, pp. 26–30, Feb. 1988

"Optical AI Architectures for Intelligent Sensors."
HESTER, C. F. and CAULFIELD, H. F.
Teledyne Brown Eng., Huntsville, AL, USA
Sponsor: SPIE
Proc. SPIE—Int. Soc. Opt. Eng. (USA) Vol. 754, pp. 185–192, 1987
CODEN: PSISDG ISSN: 0277-786X
Optical and Digital Pattern Recognition, 13–15 Jan. 1987, Los Angeles, CA, USA

"Sensor Materials for the Future: Intelligent Materials."
TAKAHASHI, K.
Dept. of Electr. & Electron. Eng., Tokyo Inst. of Technol., Japan
Sens. Actuators (Switzerland) Vol. 13, No. 1, pp. 3–10 Jan. 1988

"Development of Intelligent Sensors Using Polymer Materials."
SEO, I.
Instrumentation (Japan) Vol. 30, No. 12, pp. 32–35, Dec. 1987

"Development of Intelligent Optical Fiber Sensors."
HATTORI, H.
Instrumentation (Japan) Vol. 30, No. 12, pp. 25–28, Dec. 1987

"Distributed Piezoelectric-Polymer Active Vibration Control of a Cantilever Beam."
BAILEY, T. and HUBBARD JR., J. E.
Journal of Guidance, Control and Dynamics, Vol. 8, No. 5, 1985, pp. 605–611

"Development of Piezoelectric Technology for Applications in Control of Intelligent Structures."
CRAWLEY, E. F., DE LUIS, J., HAGOOD, N. W. and ANDERSON, E. H.
Proceedings of the 1988 American Control Conference, Atlanta, GA, Vol. 3. 1988, pp. 1890–1896

"Nonlinear Control of a Distributed System: Simulation and Experimental Results."

PLUMP, J. M., HUBBARD JR., J. E. and BAILEY

ASME Journal of Dynamic Systems, Measurement, and Control, Vol. 109, 1987, pp. 133–139

"Smart Components for Structural Vibration Control."

MILLER, S. E. and HUBBARD, JR., J.E.

Proceedings of the 1988 American Control Conference, Atlanta, GA, Vol. 3, 1988, pp. 1890–1896